move with science

energy, force, & motion

An activities-based teacher's guide

by Roy Q. Beven

Published By
National Science Teachers Association
With Support From
National Highway Traffic Safety Administration

NSTA Stock Number PB144X

Library of Congress Catalog Card Number 98-84914

ISBN 0-87355-172-9

The National Science Teachers Association is an organization of science education professionals and has as its purpose the stimulation, improvement, and coordination of science teaching and learning.

move with science: energy, force, & motion

■ CONTENTS

Introduction

Overview *v*
Curriculum design *vi*
How to use this book *ix*
From the author *x*
Acknowledgements *xi*

Notions of motion

Overview 1
ACTIVITY 1 A motion paradox 3
ACTIVITY 2 Analyzing motion 9
ACTIVITY 3 Using technology to measure motion 15
ACTIVITY 4 Describing forces that push vehicles 20
BACKGROUND READING 25

Stability when turning

Overview 33
ACTIVITY 1 A balancing act 35
ACTIVITY 2 Balanced humans 39
ACTIVITY 3 Balanced vehicles 44
ACTIVITY 4 Describing balance in turns 49
BACKGROUND READING 55

Energy and reaction

Overview 65
ACTIVITY 1 Food and fuel 67
ACTIVITY 2 Energy of motion 74
ACTIVITY 3 Energy transfers 80
ACTIVITY 4 Speeding up and stopping 86
BACKGROUND READING 88

Detection

Overview 95
ACTIVITY 1 Waves 97
ACTIVITY 2 Do you hear what I hear? 101
ACTIVITY 3 Picture this 105
ACTIVITY 4 Detection of motion 111
BACKGROUND READING 114

Collisions and safety

Overview 121
ACTIVITY 1 Collisions 123
ACTIVITY 2 Spreading forces 127
ACTIVITY 3 Dissipating energy 132
ACTIVITY 4 Investigating vehicle safety 137
BACKGROUND READING 142

Appendix

Evaluation rubric 145
Bibliography and resources 146

Introduction

■ OVERVIEW

Move with Science: Energy , Force & Motion is a secondary level activities book which uses transportation to teach the basic concepts of physics, and several areas of human biology. Specifically, this book focuses on those methods of transportation that are most familiar to high school students. The National Science Teachers Association and the Department of Transportation's National Highway Traffic Safety Administration (NHTSA) created this book with the idea that these students might learn physics more easily if they could experience it.

The learning plans described in this guide may be most applicable in an integrated science course at any level, or a conceptual physics course. It is divided into the following sections:

Curriculum Design This section explains the philosophical foundation for the development of the activities and associated instructional activities.

Activities This section is the core of the guide. Five chapters, that focus upon building ideas from the National Science Education Standards, are included. Each chapter begins with an overview of the conceptual learning plan, and contains four activities. Each activity is divided into a reproducible student's section—outlining the activity, materials, procedures and questions—and a teacher's section—describing the plan's conceptual organization and development, time management, assessment strategies, clarification of procedure, answers, and extension ideas. The five chapters are described below.

Notions of Motion A quick review of

force and motion designed to establish a basic physics vocabulary and a force analysis format; culminates with group presentations of a force analysis.

Stability when Turning Students explore stability, balance, center of gravity and rotational inertia while investigating the forces acting when an object is turning; culminates with group presentations of a force analysis of a turning vehicle.

Energy and Reaction Students investigate food, fuel, the energy of motion, and energy transfers; groups present an energy analysis of a transportation system.

Detection Students explore wave phenomena to gain a better understanding of human seeing and hearing and their importance in transportation situations; students create future vehicular detection technologies in groups, and present an analysis of their designed system.

Collisions and Safety Students investigate energy transfers during collisions to learn how safety devices work, and use their knowledge to design a safety device; students also explain scientifically founded, safe decisions about a transportation issue in a classroom presentation.

Background Readings This section contains conceptual content readings strongly connected to each activity. The individual readings are organized by subject for use by both students and teachers, and are located after the final activity in each chapter. These readings are not intended to replace a textbook, or other reference materials, but do provide accurate, relevant background information in a language similar to the instructional activities.

Resources This section provides a categorical listing of written materials, multimedia sources, technology enhancements, and Internet sites for helpful information.

■ CURRICULUM DESIGN

"Solid instructional models have students doing science and technology by engaging them in acts of exploratory investigations, constructing meaning out of their findings, proposing tentative explanations and solutions, exploring concepts again, and then evaluating concepts in reference to their own lives."
SUSAN LOUCKS-HORSLEY

Move with Science uses a constructivist learning model. Its basic science content uses a context of safety, vehicle design, and transportation as a hook for student interest. The activity sequence then uses exploratory investigations to build on that interest until students can demonstrate that they can apply their learning in new contexts. You and your students will probably find the sequence intuitive, as it builds from interest to exploration to application to knowledge transfer. You and your students will probably also find *Move with Science* a fun way to learn.

Conceptual learning plans

Learning plans are designed to provide students with opportunities to construct concepts, and each instructional activity is chosen for a purpose. Generally, while a learning plan focuses on building concepts, the activities are focused by objectives. Learning plans may be long-term organizational tools in which new activities or lessons are employed. New lessons might involve emerging technologies or new resources. Learning plans enable teachers to immediately implement new lessons into existing curriculum.

The learning plan for *Move with Science* uses a series of four activities to develop the central concepts for each chapter. These central concepts are based on content, program, and assessment goals of the National Science Education Standards. Each activity has a teachers guide and multiple extensions to help you adapt the activity to your students' needs and explore concepts further.

Attention-grabber activity This activity invites students to explore and learn while setting the stage for the other instructional activities, and establishes the concept for the chapter and the reason for further investigation. The attention-grabber activity satisfies students' need to have fun, make something, and/or talk about themselves while becoming engaged in the learning plan. The attention-grabber often starts by bringing out students' previous experiences. Students may "tell their story" while the instructor supports the discussion—being careful to establish an atmosphere of respect for each student's thoughts and experiences, and insuring an open discussion.

Real-experience activity This activity allows students to explore, discover, and construct knowledge in terms of their everyday lives. This type of activity is simple, and real enough that students could do them at home with readily available materials. As students delve into this real-experience activity, the instructor can carefully listen and pose questions to initiate metacognitive processes that may lead to deeper understanding of the concepts and systems involved. Because the interaction of the real world and the school-based topic are so tangible, the students may readily experience significant learning. This real-experience activity meets the need for students to experience their world in a safe, controlled environment instead of their daily unfocused atmosphere.

Unique experience activity The unique experience activity provides the experiences students are talking about when they say, "Guess what we did in science today?" These conversations reflect the excitement that scientists feel. The unique experience activity may be the event that steers students into careers in science, or makes science fun. It certainly should help students build an understanding of the nature of science, an area that should be revisited regularly in a science class.

Applied experience activity An applied experience activity provides time for students to propose explanations and solutions to a new problem, while using what they've experienced. In this activity they attempt to apply

Conceptual learning

Prior experiences and knowledge.

Central Idea Concepts from the *National Science Education Standards*

- **Attention grabber** A lesson to engage students in fun exploration.
- **Real experience** A lesson to explore, discover, and construct.
- **Unique experience** An ooh-ah lesson done only in a science class.
- **Applied experience** A lesson to propose explanations and solutions.

Constructed and communicated concept.

their new knowledge and skill. This should be an engaging activity, especially if the students created the "rules" earlier in the learning plan. An applied experience may be used as a performance assessment of what has been learned; however, the overall assessment of the learning plan might best be done in a cumulative portfolio in which students are asked to reflect upon all their experiences in an attempt to reveal which ones were meaningful and should be counted as significant learning experiences. Students may revisit and revise activities while explaining why they were significant to the development of the chapter's central scientific concept.

The applied experience activity allows the teacher to listen to student dialogue about the concepts learned and try to discover how (and whether) students are relating what they knew, to what they have just experienced, and to what they are doing. Applied experiences often involve a design process. As the teacher listens carefully to each group's discussion, it may be appropriate to bring everyone together to point out what is being said, and to validate or reinforce the central concept being developed.

Scientific experiment At least one of the activities in a science learning plan should be an experiment. In other words, the activity should involve a quantified, controlled, repeated, objective investigation of the effect of the manipulation of one variable upon other variables of a system. The label "scientific experiment" should only be used for activities that illustrate the nature of science. Often the real or unique experiences are scientific experiments.

Why a constructivist learning plan?

Students need to be actively involved in their science learning. But hands-on activities, while engaging, may represent little more than busy work if students are not given the time to construct the new knowledge in their minds, replacing their old, intuitive ideas with ideas consistent with the science they are experiencing. Teaching only to the objective may transmit only low-level skills, terminology, and facts to passive students. The result can be disastrous, especially for under-represented students, who often are not encouraged to rise above rote-skill learning.

Because the way students learn is not necessarily rigid, or even sequential, a learning model should not be seen as a series of discrete stages, but as an interactive process, similar to a scientific or design process. The most effective learning model for science is the constructivist model based upon the four- or five-step models that emerged around 1990. These have evolved from the learning cycle models of the early 1960s that were employed in much of the best curriculum developed in the 1970s.

Move with Science uses a four-step learning plan, and it is designed to link with the other components of instruction that enable students to learn. Combine *Move with Science* with a classroom model based upon inquiry-based learning centers to create a science course that is engaging and meaningful to all students. Focusing on the big ideas while using instructional activities that supply attention-grabbers, real experiences, unique experiences and applied experiences are lofty goals, but students are worth it.

■ HOW TO USE THIS BOOK

As part of the constructivist method, you should begin each activity in this book with a discussion of the concepts involved. Each teachers' section contains a "Conceptual Development" section which will aid you in this discussion. Whether or not your students know something about motion, or balance, and whether or not what they know is correct, they will have ideas, and should be encouraged to develop them. Have students record their original thoughts about the concepts they are learning. For some activities, you may wish to distribute background reading supplied in this book, or from another source, prior to this discussion. Either way, it is important that students have a chance to voice their thoughts, and hear those of their peers, before they begin their investigations. (Polite listening skills should be encouraged.)

Each student section contains enough information for the student to perform the activity, but does not provide a place for responses. This is because this book was designed with the use of journals in mind. A spiral-bound notebook will suffice, but a notebook that will not "encourage" students to tear pages out, such as a black-and-white essay book, is highly recommended. Journals offer students an easy way to contain their notes, an opportunity to review their original thoughts after they have progressed through a concept, and a way for you to assess their progress.

Students will maintain their journal throughout *Move With Science*, recording their thoughts both before and after an activity, as well as the results of their investigations. You may want to create a format to standardize their entries. You might want to ask your students to respond to some overall questions, like "what was your most significant learning experience?" You should also develop a rubric to evaluate the journals, and share it with students as they're putting them together. Be sure students understand that their thoughts will not be judged as right or wrong, and encourage them to record all of their ideas. You should include some questions in your assessment that no matter how students answer, they will receive full credit.

We recognize that class times vary from school to school, and have designed these activities to work within a fifty-minute class structure—*i.e.*, an activity time of two class periods will take 100 minutes. In order to fit your schedule better, questions can be assigned as homework, and extensions can be done in the classroom.

This book has been designed to be used in its entirety, but the individual activities can also be used to supplement the various lessons in your curriculum. For example, a biology teacher may not have use for an activity about waves, but an activity about how we hear or how we see may prove to be an excellent addition to his or her classroom.

Finally, because this book was designed to be used in its entirety, it has been organized to be used in order. Each chapter introduces a new concept that is integral to understanding the concept in the following chapter. Although "Collisions and Safety," the final chapter of this book, contains several exciting experiments, without a basic understanding of chapters one through four, they are only exciting examples of entertainment.

The Physics Store

Real life is very complicated yet physicists want to simplify the universe to its essence. They do this by eliminating unessential variables until they can measure the direct affect of one variable upon another. Sometimes physicists can not really eliminate unessential variables so they simply imagine an ideal situation in which ideal objects or situations exist. We call these thought experiments. For example, real-life frictional forces complicate motion so physicists often make up frictionless things. Ask students to imagine a store in which they could purchase frictionless stuff: the Physics Store. They cold purchase two frictionless skateboard from the Physics Store and use them in this learning plan. Ask the students to describe how those frictionless skateboards would affect their investigations.

■ FROM THE AUTHOR

I have come to think of the process of creating this curriculum guide as virtual teaching and professional development. I hope that the students involved with the activities of these learning plans will build important ideas and that the teachers implementing this guide will be enriched as educators.

The enrichment I've received from the many educators and friends over the past 24 years has been fundamental to the creation of this guide and I would like to recognize and thank these folks. T. Jean Adenika started me on this educational path and remains an inspiration. Many other scientists/educators have had profound effects upon me, and I've needed these teachers and people who believed in me. Thanks to all the following: Mare Taagerpera, George Miller, Frank Potter, Curt Abdouch, Frank Ireton, Ruth Von

Blum, Joan Bissell, Maureen Allen, and Bill Parker. Over the past year I've had valuable input from Gordon Chalmers and Susanne James about human biology while learning much about engineering design processes from Bob Raudebaugh. I owe a special thanks to Irwin "Sles" Slesnick, a true mentor and great educator. Finally, I could not have completed this guide without the leave and support granted me by Peter Elich, Dean of the College of Arts and Science, Western Washington University.

My personal life is very much intertwined with my professional life. I must thank the Irvine Boys, a group of vintage U.C. Irvine rowers, for all their inspiration and support—they taught me how to move. Finally I thank my wife Michele, daughter Lauren, and son Drew. I could not be an educator without them, and certainly they keep me moving.

■ ACKNOWLEDGMENTS

Roy Q. Beven, the author of *Move with Science,* is a science, mathematics, and technology educator. For two decades he has been deeply involved in all aspects of the educational systems of Orange County, California, and is currently absorbed in the creation of a new Science, Mathematics, and Technology Education Center at Western Washington University. He has won numerous awards for his exemplary classroom instruction, as well as his efforts in improving public education, including the Presidential Award for Excellence in Science and Mathematics Teaching.

As with any publication, *Move with Science: Energy, Force, and Motion,* would not be in your hands right now without the help of many people. Jeff Leaf (Virginia), Linda Kralina (Missouri), Ron Morse (New York), and Dean Zollman (Kansas) all contributed comments and ideas on the original text, and reviewed it for scientific accuracy and classroom practicality.

In addition to reviewing the text, the following classes and teachers put aside their normal routine to field test activities from the book. Many thanks to Lisa Bidelspach's Biology I Honors class at Clear Creek High School, League City, Texas; Jeffrey Leaf's periods 3, 5, and 7 TEC classes at Thomas Jefferson High School for Science & Technology, Alexandria, Virginia; Linda Kralina's General Physics classes at Mary Institute and St. Louis Country Day School, St. Louis, Missouri; Ron Morse's NYS Regents Chemistry period 10.11.AC at Minoa High School, East Syracuse, New York; and Patty Rourke's Conceptual Physics class at Potomac School, McLean, Virginia.

The staff of the National Highway Traffic Safety Administration provided review, guidance, and encouragement throughout the development of *Move with Science.* Special thanks go to Beth Poris, Jeff Michael, JoAnn Murianka, Lori Miller, Cheryl Neverman, and Ricardo Matrinez, M.D.

Move with Science: Energy, Force & Motion is published by NSTA—Gerry Wheeler, executive director; Phyllis Marcuccio, associate executive director for publications. NSTA Special Publications produced *Move with Science*—Shirley Watt Ireton, director; Chris Findlay, associate editor; Anna Maria Gillis, associate editor; Michelle Treistman, assistant editor; Christina Frasch, program assistant. Michelle Treistman was NSTA project editor for *Move with Science.* Jocelyn Lofstrom and Douglas Messier also provided assistance. Daryl Wakeley of AURAS Design designed the book and cover. Illustrations were created by Julia Sifers of Glasgow & Associates. The book was printed by United Book Press.

Cover photo: Paul Kuroda/SuperStock. Inset photo (highway sign): F. Schussler/PhotoLink. Inset photo (taxicabs): PhotoLink. Spine photo: Arthur S. Aubry/PhotoDisc.

chapter one

Notions of motion

■ TEACHER-TO-TEACHER

This is an introductory learning plan designed for you to establish a common language with your students, while reviewing some classical mechanics and establishing how to do a force analysis. For the sake of time, and the need for direct instruction, this is a teacher-centered learning plan in which students perform activities as a whole class before they are asked to present force analyses in groups. The other learning plans of *Move with Science* are much more inquiry-based and constructivist, in order to give the students time to construct new connections between physics, human biology, and transportation.

Conceptual development

Students, working on a single, whole-class investigation, learn the concept that forces always come in pairs while exploring a single push. This simple, yet profound, exploration engages students in the notion of force analysis. The following paragraphs describe the concepts developed in the four activities of this chapter.

Forces are anything that cause objects to change motion—speed up, slow down, turn, or stop. Many forces may act on an object, but they add up to one imaginary net force that can be said to act on the object. This net force is directly proportional to the size of an object's change in motion, and acts in the same direction as the object's change in motion.

Objects have a property called inertial mass. This controls how much the object changes its motion when a net force acts on it, and is inversely proportional to the object's change in motion.

Conceptual construction

Prior experiences with skateboards, ice skates, etc., as well as ideas about forces and motion.

Central Idea Force causes change in vehicle and human motion.

> **A Motion Paradox** Explore the ideas of force pairs.
> **Measuring Motion** In real life, what is the easiest way to measure motion?
> **Using Today's Technology** Measuring motion using sensors and computers.
> **Force Analysis of Vehicles** In teams, present a force analysis to the class.

Construct idea that forces cause motion and can be analyzed. Develop the skill to do a force analysis.

There are many ways to measure motion. Direct measures of speeding up and slowing down are difficult, but today's technologies make these measurements attainable. Simple measurements like distance moved are also quite useful, especially when using work or energy as concepts to describe motion.

Forces can be represented on paper as arrows or vectors. These force analyses can be used to analyze the motion of the human organism as well as on vehicles.

The motion of the human body can be described in the same manner as vehicles, with some understanding of the human biology involved. In groups, students formally present and discuss their force analyses to their peers.

National Science Education Standards

In the development of the conceptual construction of this chapter, the following Content Standard concepts and principles served as a guide.

- Formulate and revise scientific explanations and models using logic and evidence (page 175).
- Objects change their motion only when a net force is applied (page 179).
- Design and conduct scientific investigations (page 175).
- Use technology and mathematics to improve investigations and communications (page 175).
- Formulate and revise scientific explanations and models using logic and evidence (page 175).
- Whenever one object exerts a force on another, a force equal in magnitude and opposite in direction of exerted on the first object (page 180).
- The human organism has systems for movement, control and coordination that interact with one another (page 156).
- Communicate and defend a scientific argument (page 176).

Time management

The entire learning plan should take five to six fifty-minute class periods. Activities 1, 2, and 3 are each one-day investigations designed so that the whole class participates in a single analysis of one motion orchestrated by you, the teacher. Activity 4 will take two to three periods. Your students will need at least one period to prepare, and one to two for their presentations.

Learning plan assessment

Use Activity 4, or one of its extensions, as a performance assessment. Have students repeat Activity 4 using vehicles from the Physics Store (described on page x), either as a performance assessment or as practice.

Ask students to take notes on each group's force analysis, and then test the class by asking students to draw the forces in the situations presented by the groups.

Students can perform peer reviews on each presentation. Create an assessment rubric (see Appendix) before the first presentation.

Activity 1

A motion paradox

If we are both on wheels and I push you, who moves?

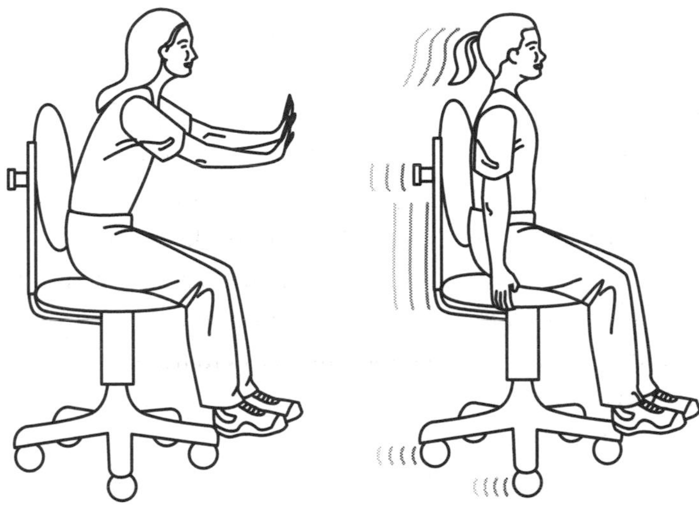

Challenge

What happens when you push off of a stationary object? A movable object? How would you measure the motion you wanted to study? What do you think you would discover?

Objectives

- To investigate force pairs.
- To describe motion in terms of measurable variables.
- To establish how to do a force analysis.

Materials

- Rolling vehicles
- Scale
- Tape measure or meter sticks

Less massive metals

Car manufacturers put a lot of resources into developing vehicles that are fuel efficient—vehicles that can get more miles to the gallon. In order to do this, scientists must take into consideration more than just the engine and the fuel, they must consider a vehicle's mass as well. In 1998, the steel industry announced the development of an ultralight steel auto body, one which has the potential to decrease the weight of a midsize sedan auto body by as much as 36 percent. The combination of new manufacturing processes and high-strength steels meets today's safety requirements, and can be made with today's technologies.

What parts of a vehicle might you modify to make it more fuel efficient? What materials would you use? What tests would you perform to determine which materials would work best?

Procedure

One Record mass of the rolling vehicle and briefly describe it.

Two Record the mass of the students involved in the activity.

Three Your teacher will lead one group of students through the following demonstration of the effects of pushing on two "vehicles." Record the data collected for each step.

a. One student will sit on each of the two rolling vehicles. Your teacher will hold one student still, while that student pushes the student on the other vehicle. The third student will measure the distance over which the push acted, and then how far the vehicle rolled. The group will repeat this several times, trying to push with equal force each time.
b. Repeat a, this time allowing both vehicles to roll. For this step, students of about the same mass will sit on the two rolling vehicles. One student will push the other, but this time the teacher won't hold on. Only one student should do the pushing; the other should just hang on to the vehicle. Repeat this several times, trying to push with equal force each time. Record the distance over which the push acted, and the distance each vehicle rolls.
c. Repeat b, this time with students of different mass on the vehicles.

Four What is the effect of the mass of the rolling student on the motion caused by the push?

Five What happens when both students are allowed to move?

Six If one force causes one object to change its motion, how do you explain a single push causing both students to move in opposite directions?

Seven Why didn't both students move when one of them was held?

Eight Decide how your class will use the following terms, and record their definitions:

- force
- mass
- force pairs

Nine Using the following guidelines, develop a force analysis of the illustration at the beginning of the activity.

- Use straight arrows to indicate force.
- The length of the arrows should represent the size of the force.
- Draw the arrow in the direction the force acts.
- Attach the tip or the tail of the arrow to the object at the places the force acts.
- Label each arrow with the name of that force.
- Draw and label a separate arrow to indicate the net force on the objects.

Ten If we are both on skateboards and I push you, who moves? A paradox is a statement that seems to be contrary to common sense, yet it is true. Explain why the answer to this is a "Motion Paradox."

■ TEACHER-TO-TEACHER

This one-period activity is built around the paradox that if one force causes one object to move, both people move when only one pushes against the other. The paradox is resolved by the assumption that all forces are really pairs of forces, and is defined in classical mechanics by Newton's Third Law of Motion—for every action, there is an equal an opposite reaction. This concept, however, is not necessarily intuitive, and many people have trouble thinking in terms of force pairs. One thing to remember is that these force pairs act on different objects, and never on the same object. This activity is an excellent opportunity to think about Newton's Third Law while analyzing force and motion.

This activity can also help you determine how much your students know about classical mechanics. It will help your class establish a common language for the discussion of force, motion, and other concepts of physics which will be discussed in later activities. Students will also learn to create a force analysis diagram, and use it as a basic tool for understanding these concepts.

You will need to choose three students of approximately the same mass to perform this activity with you. If this is not possible, then one of the students should be as close to twice the mass as possible of one of the other students. Before choosing, or asking for volunteers, remind students that their masses will be data for this activity. This may have the unfortunate result of eliminating self-conscious students from your volunteer pool, so positive encouragement may be needed.

While this group is working, the rest of the class will need to be observing quietly and recording the results. You may want to ask the class questions to prevent students from becoming distracted. If time allows, you may wish to give all students an opportunity to experience each of these three test situations. This would not be for more data collection, but rather to help them begin to connect what they experience in everyday transportation with the physics of mass and motion.

Conceptual development

- When we see an object in motion, we assume that one force acted upon it and that the object is moving in the direction of that force. However, usually more than one force acts on an object.
- These forces add up to a net force which causes an object to change its motion in the direction of that net force.
- As the inertial mass of an object increases, a given force results in less motion (see "What Is Motion?" page 25).
- When objects are free to move, we can clearly see that one force between two objects manifests as a pairs of forces, each causing one object to move—this idea is established in Newton's Third Law of Motion.

- To understand forces better, we can analyze motion using a force analysis diagram.

Materials

- Two identical, safe, rolling "vehicles": A pair of rolling chairs, or stools with backrests, are probably the safest.
- A smooth, uncarpeted floor space.
- Device to measure mass of students and rolling vehicle: A bathroom scale converted to kilograms—1 kg weighs 9.8 Newtons at sea level on Earth.
- Device to measure distance: Measuring tape or meter sticks. You should mark a starting line on the floor with masking tape so that it is easy to repeat the motion from a "zero" point.

Teacher's guide

One Ensure students understand that mass is a constant throughout this activity. Descriptions can be basic, but should include that the vehicles can move. Students should know the difference between weight and mass. You might want to review this while you are working through these first steps.

Two Same as Step one.

Three Ask the pushing student to push on the back of the chair, or firmly below the student's shoulder blades if you are using dollies or backless chairs. You may need to show them the difference between a push—full extension of the push (arm) can be measured—and a shove—full extension of the push cannot be measured. (See Extension five.) Students should push consistently. Have the third student record the distance of the push and the roll, and then roll the student and chair back to the starting point.

a. Anchor the student by pressing down on his/her shoulders. If you do not anchor the chair well, it may move, which will ruin the integrity of the activity.
b. Have two students of approximately equal mass sit on each rolling vehicle, and ask the pushing student to push with the same force as the first time. Do not hold anyone. Both students should roll equally, and both vehicles should roll half as far as in Step three a.
c. Repeat Step three b, with students of different mass on the vehicles. If one student is approximately twice as massive as the other other student, their vehicle will roll only about half as far as the lighter student's. (If you have three student of about the same mass, you could have two sit on one vehicle and one on the other.)

Four As mass increases, motion decreases, if force remains constant. (See Extensions one and three.)

Five Both students move even though there is only one force.

Six There must have been two forces.

Seven Answers to this question will vary. Students should definitely state that you prevented the student from moving, but may stop after explaining that you were holding the student. Students should understand that by holding on to the student, you "attached" the student to the Earth, thus adding its mass to the student's. The Earth has a mass of about 60 x 10^{22} kg, while the student/vehicle may have had a mass of only 60 kg. The Earth is 10^{22} (10 billion trillion) times as massive as the student/vehicle, and relative to that object it does not move. The Earth did receive a force, but the resultant motion is smaller than we can measure, therefore effectively zero.

Eight Students should record these definitions in their journal, and you might want to leave them on the board for the remainder of this chapter. The language you use is very important. Be certain that all of the students understand the definitions, and can use the terms correctly. (See Extension two.)

Nine Diagrams will vary. Ensure students understand how a force analysis should be constructed. Figures 1.1 and 1.2 show force analyses of the activity before and after the push.

Ten Students should state that both people will move. They may go further, and define this as an action-reaction situation. They may also identify their new understanding of force pairs as the cause of the action-reaction motion. The paradox is that one force causes two separate motions where we might assume it should only cause one.

Extensions

Extension one Some students may want to measure time in addition to, or instead of, distance. If this is the case, encourage them to do so. Have them record and compare stopping times to the mass of the rolling student and the distance traveled, and hypothesize about the relationship between these variables.

Extension two Review the concept of force pairs by discussing activities such as ice skating, and skateboarding in terms of force. If an ice skater applies force to move forward, what other objects around the skater are affected? For extra credit, ask students to investigate their motion on a skateboard, or other self-propelled vehicle, and explain it in terms of force, force pairs, mass, and distance.

Extension three Explain to your students that a car with an automatic transmission will move by itself, on a flat surface, if put in drive. Ask them what they could do to that car—other than putting it in park, stepping on the brake, or physically restraining it—to keep it from moving forward?

Extension four Have students compare pushes to automobile engines. Why does a large vehicle need a larger engine than a smaller vehicle? What happens when a large engine is put in a small car?

Extension five Distribute an enlarged photocopy illustration of the human arm (Figure 1.3). Have students draw the muscles of the upper arm in both positions. Ask students to explain how their arm moves in one direction while two muscles are moving in opposite directions. (You may need to provide some anatomy resources.)

FIGURE 1.3
Arm pushing.

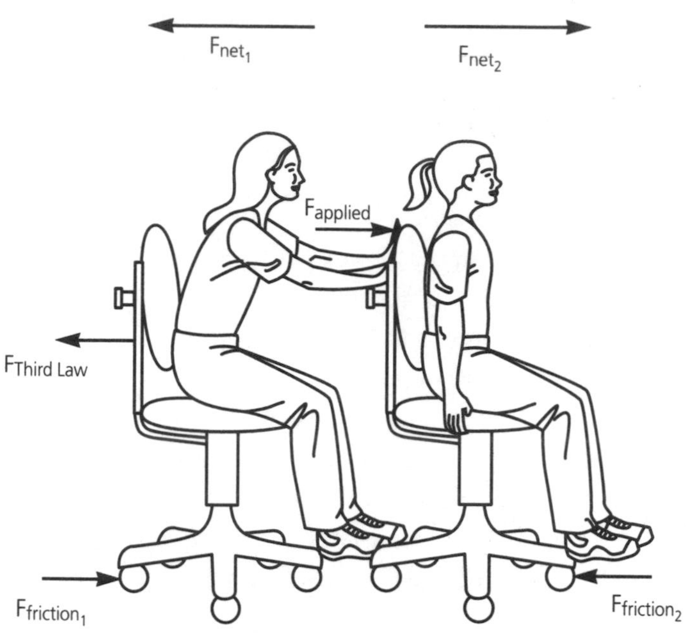

FIGURE 1.1
Force analysis diagram before the push.

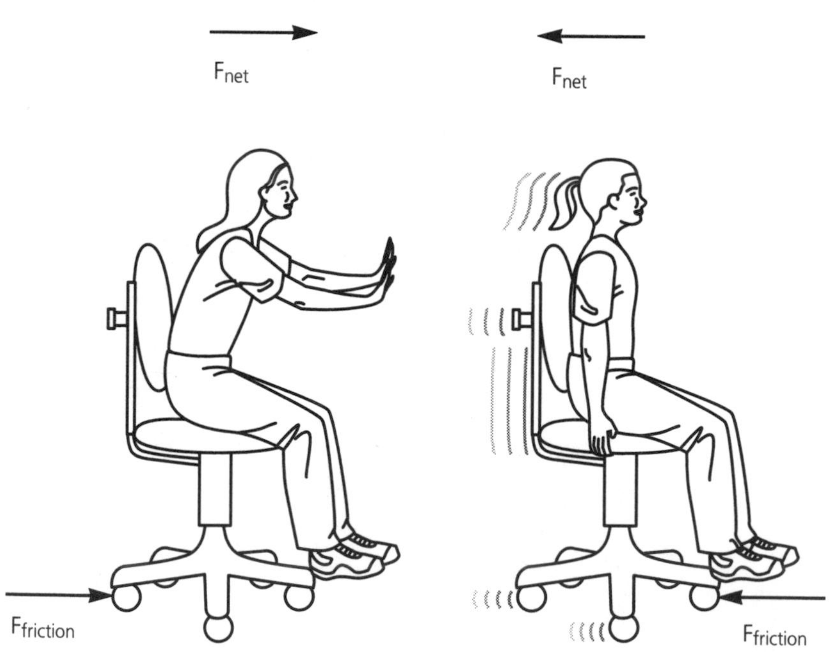

FIGURE 1.2
Force analysis diagram after the push.

Activity 2

Analyzing motion

When I push you, who moves more?

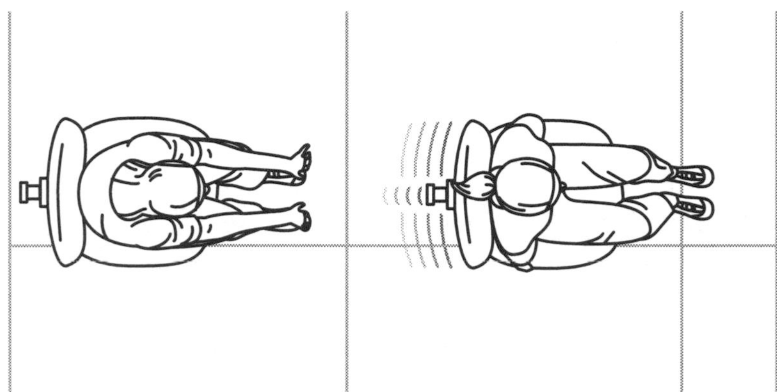

Challenge

How well can you measure force and motion, and the effect of force on motion? In studying force and motion, what variables can you control, and how? How could you measure a push and its resulting motion accurately? What might this quantitative data tell you?

Objectives

- To control and measure a force or push from a human arm.
- To control and measure the distance over which the pushing force acts.
- To measure the resulting motion and distance traveled caused by a force.
- To determine relationships between force and motion based upon the data collected.

Materials

- Rolling vehicles
- Scale
- Meter stick or tape measure

Procedure

One As a class, devise a strategy for analyzing motion.

Two What variable(s) can you control in your analysis? How? What can you not control? Why not?

Three Explain the problems you think you may encounter while trying to control the variables in this investigation.

Four Your teacher will utilize the strategy you developed in step one, while leading a group of students through the investigation.

Five Create a separate data table for weight and mass, force, and distance, and record the following data:

- Weight and mass of students
- Weight and mass of rolling vehicles
- Force (measured in Newtons) applied in each trial
- Distance the force acted over in each trial
- Distance traveled in each trial

Six What patterns do you see in the data?

Seven Summarize the effect of one push on the motion of the two students. Be certain to use only the data you have collected. Try not to draw upon ideas or experiences which were not observed in this activity.

Eight What future investigation might you perform in order to draw more conclusions about force and motion? (See Extension three.)

Nine When I push you, who moves more?

■ TEACHER-TO-TEACHER

This activity is designed to establish the use of data and the process of scientific investigation using the human organism. It will also challenge your students' understanding of the basic concepts of force by having them develop a strategy for analyzing motion. The purpose is not to verify a formula, but rather to establish relationships quantitatively.

Motion is tricky to measure. Students will find it difficult to apply the same pushing force consistently. Using distance rolled to measure motion may be an intuitive method for your students, but it is not a common scientific choice. Keep students focused on the process of controlling variables by manipulating one and measuring the results. After collecting data, help them use inferential reasoning to draw conclusions and design further investigations.

The procedure for this activity is similar to that of Activity 1, and will take one period. If you do not use "A Motion Paradox" in your class, you may want to photocopy Step three (page 4) for your students to be used as a guideline, rather than as the actual strategy.

Conceptual development

- Science requires measurable quantities, controlled processes, and inferential reasoning.
- To "analyze scientifically" means using a scientific process.
- Using a scientific process to analyze force, motion, and the human organism is difficult.
- Scientific data can be obtained through scientific process, and interesting conclusions can be inferred about the relationship between force and motion.

Materials

- Two identical, safe, rolling "vehicles": A pair of rolling chairs, or stools with backrests, are probably the safest. (Same as Activity 1.)
- A smooth, uncarpeted floor space. (Same as Activity 1.)
- Device to measure mass of students and rolling vehicle: A bathroom scale converted to kilograms—1 kg weighs 9.8 Newtons at sea level on Earth. (Same as Activity 1.)
- Device to measure distance: Measuring tape or meter sticks. You should mark a starting line on the floor with masking tape so that it is easy to repeat the motion from a "zero" point. (Same as Activity 1.)

Teacher's guide

One If your class has gone through Activity 1, then a strategy may be devised easily. If not, encourage creativity, but be aware of time. Spend some time discussing the force of a push. Will it act over the same distance each time? How do you measure it while pushing? Should you double or triple it? Students may see that, although measuring force while pushing is not easy, they can use the human ability to "feel" the amount of push based on a practiced standard. Perhaps a group would like to practice pushing consistently with the "pusher." The distance over which the force acts must be constant. The reason for this, work, will be explored in Chapter 3, "Energy and Reaction" (page 65).

Two Students can control the force they apply in this investigation by controlling their "push." They cannot control the mass of the individual students and the rolling vehicle, but the overall mass can be increased by having a second student sit on the vehicle. If necessary, review the difference between weight and mass.

Three Students should conclude that they will have problems controlling and measuring force. Humans are not perfect machines, and cannot apply the exact same force for each trial.

Four Be sure the investigation procedure is feasible, and that you have all the materials that are necessary. Select three students who are able to follow directions well; see "A Motion Paradox" (page 3). (See Extension one.)

Five Student data tables will vary. See Figure 1.4.

Six Several patterns will appear if the students are consistent. Patterns will depend upon which variables were changed and which were kept constant. For example, greater force applied, equals greater distance traveled. Greater mass, however, equals less distance traveled, if the force applied is the same. Some students might then infer that a greater mass requires more force to cover the same distance. (See Extension two.)

Seven Answers will vary, but should show an understanding of the basic concepts learned in "A Motion Paradox" (page 3).

Eight Students may want to have better measuring tools (see Activity 3, page 15), or may want to manipulate other variables such as friction, by changing the surface of the floor or the wheel mechanism. (See Extension four.)

Nine A less massive person will move more as a result of a push, or applied force, if all other variables are equal. (See Extension three.)

Extensions

Extension one If time allows, encourage your students to investigate further the relationship between force, motion, and mass, by staging additional trials with widely varying masses on each rolling vehicle.

Extension two Ask students how they think stopping time and distance is affected by force and mass. Have students test their hypotheses by performing the following investigation. Attach some rope to the rolling vehicles so that they can be pulled. Have one student pull another who is sitting on the rolling vehicle, making sure the "puller" is able to move away from the front of the vehicle when he/she lets go. A relatively equal force can be applied if the student/vehicle is simply dragged, not yanked, along to a marked point on the floor, and let go.

Students can measure stopping time using the second hand of a watch or the classroom clock, and should start timing as soon as the pullers let go. Repeat several times with different masses. Students should find greater stopping times and distances for greater masses, but not twice the time and distance for twice the mass. Ask students to relate their data to traffic laws and speed limits, and different size vehicles such as sports cars and large trucks.

Extension three What variables, other than those identified in Step two, might make the situation described in Step nine unequal.

Extension four Choose several "future investigations" from Step eight and have your students work through them. Or, as a take-home assignment, have each student create an activity, based on their responses to step eight that the class can do. Each activity should include a challenge question, objectives, materials, a procedure, questions, and answers.

DATA FOR MASS

	PERSON 1	CHAIR 1	TOTAL #1	PERSON 2	CHAIR 2	TOTAL #2
MASS (Kilograms)	50 kg	15 kg	65 kg	50 kg	15 kg	65 kg
WEIGHT (Newtons)	490 N	147 N	637 N	490 N	147 N	637 N

DATA FOR MOTION: Person 1

	TRIAL 1	TRIAL 2	TRIAL 3	TRIAL 4	AVERAGE
TOTAL DISTANCE TRAVELED (meters)	0.80 m	0.58 m	0.85 m	0.97 m	0.80 m
DISTANCE FORCE ACTED (meters)	0.30 m	0.30 m	0.30 m	0.30 m	0.30 m
DISTANCE CAUSE BY FORCE (meters)	0.50 m	0.28 m	0.55 m	0.67 m	0.50 m

DATA FOR MOTION: Person 2

	TRIAL 1	TRIAL 2	TRIAL 3	TRIAL 4	AVERAGE
TOTAL DISTANCE TRAVELED (meters)	0.90 m	0.87 m	0.91 m	0.92 m	0.90 m
DISTANCE FORCE ACTED (meters)	0.30 m	0.30 m	0.30 m	0.30 m	0.30 m
DISTANCE CAUSE BY FORCE (meters)	0.60 m	0.57 m	0.61 m	0.62 m	0.60 m

FIGURE 1.4
Sample tables and data from Step five.

Activity 3

Using technology to measure motion

How can technology help us analyze motion?

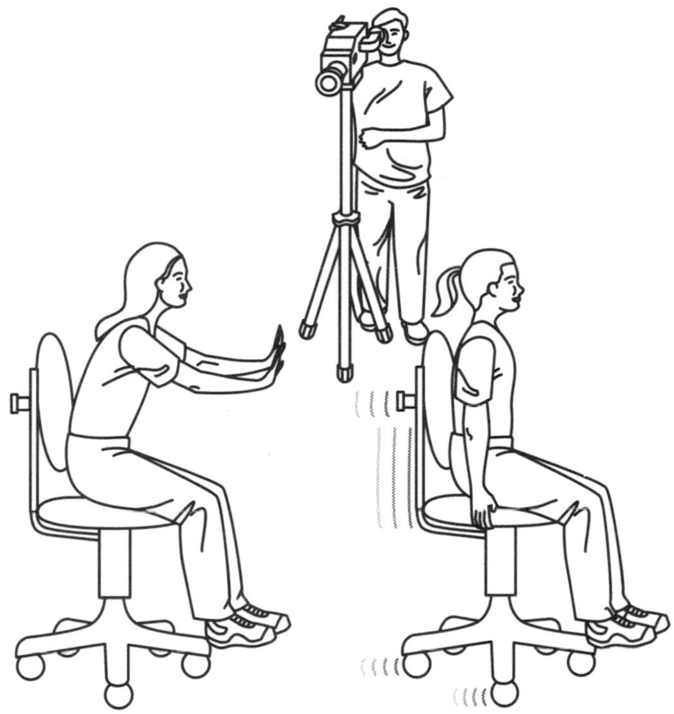

Challenge

What might you discover when you use technology to record motion? Can technology increase scientists' precision? Accuracy? Both?

Objectives

- To control and measure a force exerted by a human arm
- To control and measure the distance over which the force acts
- To measure the motion caused by a force
- To determine a relationship between force and motion based upon the data collected

Materials

- Rolling vehicles
- Scale
- Tape measure or meter sticks
- Technologically-enhanced measuring devices for force, distance, mass and/or weight

Procedure

One Your class will use technology to examine the effect of force on motion. Your instructor will list the various technologies available to you, and lead one group through the strategy you created in Activity 2. Describe the available technologies.

Two What is precision? Can technology increase the precision of your investigation? If so, how?

Three What is accuracy? Can technology increase the accuracy of your investigation? If so, how?

Four What variables can you now measure with greater precision?

Five Develop a data table that reflects the technology you used in this activity. Investigate the effect of varying the force of a push on the resulting motion.

Six Summarize the data by graphing each person's motion.

Seven What can you infer about force and motion from your data?

Eight How can technology help us analyze motion?

■ TEACHER-TO-TEACHER

This activity allows students to explore the relationship between force and motion using technology. Repeat the procedures you followed in Activity 2. You may want to review the concept of speed, and how it is a measure of motion, with your class. (See "What is Motion?" page 25.) The activity will take one period.

Conceptual development

- We have senses that enable us to analyze certain aspects of force and motion.
- To enhance our sensory perceptions of the universe, we have developed—and continue to develop—new technologies.
- Objects in motion are difficult to measure but, by recording their motion with technological devices, we are able to analyze motion more precisely.
- With technology, we can repeat motion analyses, thus allowing us to improve accuracy.

Materials

- Two identical, safe, rolling "vehicles": A pair of rolling chairs, or stools with backrests, are probably the safest. (Same as Activity 1.)
- A smooth, uncarpeted floor space. (Same as Activity 1.)
- Device to measure mass of students and rolling vehicle: A bathroom scale converted to kilograms—1 kg weighs 9.8 Newtons at sea level on Earth. (Same as Activity 1.)
- Device to measure distance: Measuring tape or meter sticks. You should mark a starting line on the floor with masking tape so that it is easy to repeat the motion from a "zero" point. (Same as Activity 1.)

Possible technological devices

- Video Analysis

  Equipment Video camera, video tape, lights, good VCR (four-head with frame-by-frame stop motion), and video monitor.

  Procedure To establish scale, record the experiment from the side with a meter stick in the field of view. Review the video using single frames. The industry standard is 30 frames a second, therefore there is $\frac{1}{30}$ of a second time interval between frames. Students can determine time by counting the frames it takes for the student/vehicle to travel the distance.

- Computer-Based Lab (CBL) or Probeware

  Equipment Two distance probes, interface, computer, monitor, printer. There are many inexpensive distance probes available that can be used with modern or older computers.

  Procedure Set up the two distance probes to monitor each student's motion. There are also force probes available which can monitor the push.

- Photogate Timers
  Equipment Two photogate timers and appropriate "flags" for each student.
  Procedure Set up a photogate timer on each person. These can record the time interval for the "flag" to pass right after the push. This time can be used to calculate the speed at the point knowing the width of the "flag."

Teacher's guide

One Descriptions will vary depending upon which equipment you use for this activity, but all should include an understanding of how the equipment will aid in the analysis of force and motion—a video camera will record motion and allow the viewer to study the recording versus having to repeat the experiment. Lead one group through the steps of Activity 2 (page 9) using the chosen equipment. If more than one device is available, and time allows, repeat the experiment with other devices and compare the results. (See Extension one.)

Two Precision refers to the exactness of a measurement, or how well equipment can measure. It is increased because small time intervals and distance can be recorded and reviewed.

Three Accuracy refers to the degree of how "error free" a measurement is, or how well a scientific group can measure. It may or may not be increased. If the students are sloppy, they will have nice, precise, inaccurate measurements. If time allows, encourage a discussion about what technologies exist, or could be created, to increase accuracy in this activity and elsewhere.

Four Students should note that both distances and time intervals can be measured with more precision. If you use force probes, then force can be measured with more precision.

Five Data tables will vary.

Six Because of the more precise data, student summaries should be more complex in this activity than in previous ones. Graphs should show a clear relationship between force— on the horizontal axis—and motion—on the vertical axis—and students should base their conclusions on their own graphs. A CBL set-up can graph the data for the students. (See Extension two.)

Seven Equal forces acting on equal masses over the same distance should result in the same speeds or motion. Less massive objects will have more speed, which is exponentially (to the second power) proportional to their mass. Students may not have enough data to infer the exponential nature of this relationship, however. They may simply say less mass equals more speed, or more mass equals less speed.

Eight Students should note that technology gives them the ability to record and/or analyze precise data, which is very helpful for studying motion.

Extensions

Extension one Ask students to consider how the distance probe technology can be applied to the human body. (The distance probes come from the autofocus component of cameras. It is related to the use of piezoelectric crystals to create and pick up high pitched sound, and is used in ultrasounds.)

Extension two Have students graph the relationship between mass (x-axis) and motion (y-axis), with the understanding that force must be equal for each mass pushed.

Activity 4

Describing forces that push vehicles

Does everything push against something? Do both objects always move? What do vehicles push against?

Challenge

Have you ever tried to describe the forces that cause the motion of vehicles? How does pushing on a bicycle pedal result in forward motion? How can an engine push a vehicle forward? If you did a force analysis of these situations, could you explain it to your peers?

Objectives

- To use arrows or vectors to describe forces that cause vehicles to move
- To create a force analysis for a specific scenario and present it to the class

Materials

- Group force analysis packet
- Transparency sheets
- Transparency pens

Procedure

One Your class will be divided into groups, each of which will be given a motion scenario. Describe the forces that cause the vehicle in your group's scenario to move. Do not write on the group summary sheet yet.

Two On your individual force analysis sheet, sketch each of the forces you think are acting on the vehicle. Use the guidelines from Step five below.

Three Share your force analysis sheets with the rest of your group. Record the differences. Did you forget, or incorrectly add a force?

Four As a group discuss the merits of each member's ideas. Remember to respect each other's thoughts!

Five In pencil, start to develop a force analysis on your group summary sheet. Use the following guidelines to make your force arrows:

- Use straight arrows to indicate force.
- The length of the arrows should represent the size of the force.
- Draw the arrow in the direction the force acts.
- Attach the tip or the tail of the arrow to the object at the places the force acts.
- Label each arrow with the name of that force.
- Draw and label a separate arrow to indicate the net force on the objects.

Six Ask your teacher to review your force analysis.

Seven Finish writing out the force analysis on your group summary sheet.

Eight Individually, answer the questions on your individual force analysis sheet.

Nine As a group, prepare your presentation outline.

Ten Make a transparency by tracing your group's force analysis onto a clear transparency sheet. Use only the special transparency pens provided.

Eleven Practice your group presentation and then present your force analysis to the class.

Have force will travel

In order for your bicycle to move, you must apply a force. No peddling means no force, and no force means no motion. But, what if you didn't have to pedal all the time?

In the 1998, ZAP Power Systems opened the United States' first electric bike store. The 10 kg ZAP Power System can be added to any bicycle to give it a "power assist" when needed, and reduces emissions 99 percent over gas vehicles. The system can be activated or deactivated with the flick of a switch, allowing the rider to pedal with or without the system. The system can even regenerate energy while the bike is freewheeling and store it for later use.

Would the additional mass of this system affect the force needed to move the bicycle? How would a force analysis of an electric bicycle be different from that of a normal bicycle?

■ INDIVIDUAL FORCE ANALYSIS SHEET

Procedure

One Sketch the forces you think are acting on the vehicle.

Two Share your thoughts with your group.

Three Compare and contrast your individual force analysis with your group's final analysis. Record the differences and similarities.

Four How much difference is there between the size of the applied force and the net force on the vehicle? Why?

Five What role does the rider/skater/driver play in your situation?

Six How is this situation similar to the situation in which one student pushes off another? How is it different?

Seven Does everything push against something? Do both things always move? What do vehicles push against?

■ TEACHER-TO-TEACHER

This activity asks students to employ the basic concepts of force and motion in a force analysis of a real transportation situation. You can choose any or all of the motion scenarios from the eight provided (pages 29-32), or create your own. Base your choice on student experience and ability. If necessary, replace the downhill situations with horizontal motions, by tilting the original before photocopying, to simplify the force analysis.

This activity also introduces students to the idea of defending their work, by presenting their findings in front of their peers. Each student must participate in their group's presentation, and groups should be encouraged to outline their presentation, noting who will speak for which section. Students should take notes on the presentations, and each presentation must end with a question/answer time. This will help establish that students should listen to each other, as well as encourage each student to feel that their hypothesis is important. The presentations are intended to be a successful experience and not a test, although you can use it as an assessment tool. Student preparation will take one period, and presentations will take one to two periods.

Materials

- Group Force Analysis Packets: One motion scenario, with enough Individual Force Analysis Sheets for each group member.
- Clear transparency sheets and colored transparency pens: Distribute one sheet and pen to each group after you have approved their analysis and the groups are ready to prepare their presentation.

Teacher's guide

(Individual Group Packets)

One Individual sketches will vary for each student, but should follow the guidelines in the procedural section.

Two Remind students that they are to complete this step individually. Be sure to monitor your available time.

Three Students should write about how their individual force analysis compared to the group's final analysis.

Four Student should state a force, like "the push on the bike pedal." They should then explain how there are many forces that go into this particular applied force, and that the forces are simplified into one net force.

Five The net force on the vehicle is the ground, water, or air pushing the vehicle forward. It is probably much smaller than the force applied by the rider. Some of the rider's force results in energy converted into heat and "lost" through friction forces acting on the vehicle. The rider/skater/driver is the source of the work (energy) and/or the controller of the motion.

Six The similarity will always be the idea of pushing off of something, and the differences will vary with different situations.

Seven This is an opportunity for you to be as specific or general as you wish, depending upon the progress of the class and the time remaining. A discussion could cover topics such as air having mass and that air molecules push against each other. An example of this can be demonstrated in the classroom by having a student blow against a piece of paper, or by having a student cause a piece of paper to slide off of a desk just by walking by it. Students should conclude that vehicles push against each and every thing it comes into contact with, especially the road on which it is traveling and the passengers who travel in it.

Extensions

Extension one As an extra credit assignment, have students create a force analysis of another motion scenario, different from the one they worked with in class. They can use a scenario from this book, one they create, or an illustration from a newspaper or magazine.

Extension two Have students research force within the body. This activity can provide a link to some basic physiological processes, such as how the contraction and relaxation of muscles in the leg can result in the forward motion of a bicycle. A force analysis can be drawn of the force and motion of the leg walking or riding, or of an arm throwing or paddling. This extension will require students to research the muscles of the leg, or arm, and how they interact. They should focus solely on muscles, and not tendons, nerves, and bones.

Background reading

What is motion?

The scientific community debated this question a few hundred years ago and came up with some answers that are now part of the theory we call **classical mechanics**. A couple of thousand years ago, however, the scientific community had some different ways of looking at motion which we call **ancient physics**. Today, in **modern physics**, we use a few theories to explain motion. We use **classical mechanics** for almost every motion in our daily lives. We use **special relativity** to explain motion close to the speed of light, and we use **quantum mechanics** to explain motion at the atomic level. Clearly "What is Motion?" is not a trivial question.

According to ancient physics, the natural motion of an object was for it to be stationary. Every object had a natural position, and to move an object out of its natural position you needed to come into contact with it to apply a force. This theory also stated that objects naturally moved back to their natural position without the aid of an applied force. For example: A rock was made of earth, its natural position was on the ground, its natural motion was stationary. You could lift the rock by applying a force and when you let it go it traveled naturally back to the ground—its natural position—and stopped there.

One main difference between ancient physics and classical mechanics is this idea of **natural motion**. In classical mechanics, we define a characteristic of objects called **inertia**. This is the property of an object that controls its motion, and we measure it in kilograms. It is the inertia, or **mass** of an object, that determines how much an object changes its motion when a particular force is applied. The intertia of an object will keep an object in motion forever, or at rest forever, unless a net force is applied to it. However, inertia itself cannot be overcome or canceled, just like mass cannot be created nor destroyed. Objects simply have inertia, and that is what controls their motion. So, the natural motion of an object is whatever it is doing when no force is applied, whether the object is stationary or traveling at 100 kph. You might ask "why do objects have inertia?" Science is still looking for the answer to that question.

You might also be wondering if a fast-moving object has a lot of motion. The answer is, not necessarily. We should all agree that a massive object, such as a large truck or a football lineman, has more motion than a less massive object, like a compact car or a figure skater, when the two objects are moving at the same speed. We probably think this because we know a greater force is needed to accelerate these objects of different mass—and we have seen the results of collisions between some of these objects.

We can measure many factors of motion—distance, time, speed, acceleration, direction—but let's look at speeding up an object to determine how much motion it has. In the activities of "Notions of Motion," you applied a force over a distance to **speed** objects up. You did **work**, the product of a force acting over a distance which is measured in Newton-meters.

$$W = fd$$

One Newton-meter is defined as one Joule, our basic unit of energy. A way of measuring motion is to define the energy an object has solely because of its motion. This concept is called **kinetic energy**. (Work and kinetic energy will be covered in Chapter 3, "Energy and Reaction," page 65.)

In the activities of "Notions of Motion," two students received the same force acting over the same distance, essentially the same work. Therefore, both students would have the same kinetic energy, or motion. However, this does not necessarily mean they had the same speed. Kinetic energy, as well as speed, is related to mass, and, as you have discovered, a more massive object needs a greater force to speed it up over the same distance. Therefore, if one student is more massive than the other, that student would not have as much speed as the less massive student, but may have the same motion.

When we explore force, distance, speed, and mass, we see an interesting relationship emerge between motion, kinetic energy and speed. As you apply a force on an object, it speeds up or accelerates: twice the force results in twice the acceleration, or change in speed. But when a force acts over twice a distance, the object does not speed up to twice as much. To really understand this requires some immersion in quantitative analysis employing algebra and/or calculus. The result, however, is that twice the work results in twice the kinetic energy, but only 1.4—the square root of two—times as much speed. Conversely, if you double the speed of an object, its kinetic energy increases by four—the square of two. Motion, as measured by kinetic energy, is exponentially—power of two—related to speed. Motion is directly related to mass, therefore motion or kinetic energy is related to the product of mass and speed squared.

$$KE = \frac{1}{2}mv^2$$

There are other ways of measuring motion, such as **momentum** and **angular momentum**. Momentum is the quantity of motion as measured by the product of mass and speed considering direction. Angular momentum is the amount of spin as measured by the product of rotational inertia and angular speed with a direction parallel to the axis of rotation. Although the "Stability When Turning" section of this book imparts a basic understanding of angular momentum, these exciting concepts are best covered in a more advanced physics course.

Force, motion, and force pairs

Force and constant motion In science we often describe single forces that cause a change in motion of a single object. In everyday life we know that we need a force just to keep moving, as well as an additional force to change motion—speed up, slow down, or turn. The force needed to keep an object moving is the force that cancels the frictional forces acting to slow the vehicle and ultimately stop it.

An object like a planet does not need a force to "push" it along because there is no atmosphere in space, and therefore no friction causing surfaces. A vehicle on a freeway traveling at a constant speed in a straight line—not speeding up, slowing down or turning—also needs no net force to keep it moving, but for a different reason. The force supplied by the vehicle's engine goes toward canceling the frictional forces that would stop the vehicle if the engine stopped. To check a force analysis like the ones in "Notions of Motion," simply ask yourself if the vehicle is speeding up, slowing down or turning. If not, then the net force is zero. If so, then there is a net force acting upon the vehicle.

Turning and stopping In space, the gravitational force between a planet and the sun is the force needed to turn the planet which keeps it in orbit. The planet will stay in that orbit forever if no other forces act upon it. In a vehicle, the force between the tires and the road, friction, is the force that may turn the vehicle when the wheels are turned. If there was no friction, a vehicle could not turn. Friction is reduced when something comes between the tires and the road, such as water or ice, or when the tires become too smooth. It is the frictional force that stops a vehicle as well.

Where the rubber meets the road
So, a vehicle's tires must have friction to speed up, slow down, or turn. The greater the tire's contact with the road, the greater the friction. Many racing cars use soft, wide, slick tires for maximum friction to speed up and maneuver quickly. However, a little drop of water would spread evenly under those tires resulting in hydroplaning, a very dangerous situation. Therefore, everyday tires are treaded. Treaded tires give water a path away from the place where the tires contact the road, while also providing surface area to generate enough friction to turn and stop the vehicle. The same is true for all vehicles which use rubber tires.

Force pairs A vehicle's tires, like our feet, push against the ground to create the force that causes the vehicle's motion. Everything must push against something! One thing to remember is that force pairs never act on the same object so they may never simply cancel themselves.

Does that mean Earth is being pushed a

bit by every vehicle, person or animal pushing against it to move? Yes, because all forces come in force pairs. When you push against the ground to walk or run forward, Earth receives a "slight" push backwards. Remember Earth is 600 billion trillion kilograms (6 x 10^{22} kg), so the resultant motion from all our pushes is very, very, very little. A couple of classic physics thought experiments related to this are as follows:

- If all the people on the planet were to walk toward the west, what would happen to the length of the day?
- If all the people in China were to simultaneously jump up, what would happen to Earth?
- If you repeated the activities in "Notions of Motion" on chairs in outer space. What would it look like? How would the conditions in outer space affect your experiment?
- If you played catch in outerspace, what would it look like? How would the force of the throw affect the thrower, and the force of the ball affect the catcher?

chapter two

Stability when turning

■ TEACHER-TO-TEACHER

This learning plan is designed to provide your students a significant amount of time to understand the somewhat difficult, but interesting, concepts described below. Students are "hooked" by a fun balancing activity, and develop a definition of stability as they explore how humans and vehicles maintain balance. They complete the learning plan by presenting a force and stability analysis to the class. This learning plan builds on the force analysis skills and presentation experience of "Notions of Motion" (page 1).

Conceptual development

Students explore and describe how they balance objects in their hands while manipulating the distribution of the mass of the object. They employ scientific words—base, center of gravity, rotation, rotational mass—in their descriptions. The following paragraphs describe the concepts developed in the four activities of this chapter.

Students examine the relationship between their own center of gravity and stability. They discover the effect of closing their eyes on their ability to balance, as well as how they react to different distributions of their own mass.

Students investigate the relationship between the angle of a vehicle and its stability in static and dynamic situations. They discover the effect of the distribution of mass in a truck, and practice doing force analyses on turning vehicles. Objects have a property that controls how they rotate called rotational mass, spinning inertia, or moment of inertia. An object's rotational mass can be significantly different from its linear mass.

Conceptual construction

Prior experiences of balancing when standing and moving.

Central Idea Force acting can make a person or vehicle unstable.

- **A Balancing Act** Explore balancing objects with the hand.
- **Balanced Humans** Explore how humans balance in various positions.
- **Balanced Vehicles** Investigate the effects of roadway banking.
- **Describing Balance** In teams, present a force analysis to the class.

Constructed and Communicated Idea Forces need to be balanced for a stable vehicle.

Students, in groups, formally present and discuss their force analysis of an object's stability when turning to their peers.

National Science Education Standards

In the development of the conceptual construction of this chapter, the following Content Standard concepts and principles served as a guide.

- Formulate and revise scientific explanations and models using logic and evidence (page 175).
- The human organism has systems for movement, control, and coordination that interact with one another (page 156).
- If more than one force acts on an object along a straight line, then the forces will reinforce or cancel one another, depending upon their direction and magnitude. Unbalanced forces will cause changes in the speed or direction of an object's motion (page 154).
- Communicate and defend a scientific argument (page176).

Time management

This learning plan may take as many as 10 or 11 fifty-minute periods, or as few as seven, depending upon how much you and your students delve into the concepts and their applications. This plan could be implemented nicely in a distinct time period, like between Thanksgiving and winter break. Activities 1 and 2 can take one to two periods. Activity 3 is a very engaging investigation that, with discussion time, will probably take three class periods. Activity 4 will take at least two periods for students to prepare, and one to two periods for their presentations.

Learning plan assessment

Use Activity 4, either as written or with vehicles from the Physics Store, as a performance assessment.

Ask students to take notes on each group's force analysis, and then test the class individually by asking students to draw the forces in each situation presented by the groups.

Students can perform peer reviews on each presentation. Create an assessment rubric (see Appendix, Figure A-1) before the first presentation.

Activity 1

A balancing act

How do you balance an object?

Challenge

Have you ever tried to balance an object that appeared much too hard to balance? What did you do? Can you scientifically explain how performers achieve their balancing acts? How about balancing toys?

Objectives

- To observe and describe balanced objects in terms of their base, center of gravity, rotation, and rotational mass.
- To describe the stability of a balanced object.

Materials

- Meter stick
- Clamp
- Metal rod

Procedure

Note: In this activity you will be balancing long, heavy objects. Establish a sufficient area where your group can work, and use caution while working. Do not walk into other groups' areas, and beware of falling objects!

One Try to balance a meter stick vertically on the palm of your hand. Watch the other students in your group try. Describe how you and your group kept the meter stick from falling. Use the words base, center of gravity, and rotation.

Two What are the forces acting on the meter stick when it is balanced? What is happening when it tips over?

Three Balance the meter stick horizontally on your hand. Is it easier, or more difficult, to balance horizontally or vertically? Why? Without holding onto the stick, how can you keep it from rotating while on your hand?

Four Use the "C" clamp to manipulate rotational mass by attaching it at different locations on the stick. Start with the clamp at 50 cm. Try to balance the meter stick upright on your hand. How does the clamp's large mass affect your ability to balance the stick?

Five Will the stick be easier to balance with the clamp close to its base or further away? Before moving the clamp, each member of your group should write down what they think will happen.

Six Move the clamp to 10 cm from one end of the meter stick. Balance the stick with the clamp close to, and then away from, its base. Describe your results in terms of the object's base, center of gravity, rotational mass, and rotation. Did you guess correctly?

Seven Find the center of gravity of a straight metal rod, and balance the rod flat across your index finger. How stable is the balanced metal rod?

Eight After everyone has balanced the metal rod, have one group member bend it at the balance point to an angle of about 145°. Each member of your group should now balance the bent rod on their index finger. Describe the stability of the rod. Why does the rod behave differently when it is bent?

Nine How do you balance an object?

TEACHER-TO-TEACHER

Your students have probably tried to balance at least one thing in their lives, and this activity provides an exciting link between curiosity and science. This activity will take one to two periods, and you can set the stage for it by discussing their exploits, whether as formal as performing on a balance beam, or as simple as trying to balance an object on their hand. Hopefully, they will include gravity as a factor in their explanations. If not, help them connect their discoveries about force and mass from the Notions of Motion activities with balance. Prior to beginning this activity, you should discuss the terms "balance," "center of gravity," and "base" with your students.

In addition to their own experiences, students may be familiar with several balancing toys, such as the small toy bicycle ridden by a stuffed animal that easily stays on a tight rope, or the plastic bird that can balance on the tip of its beak on anything. In each situation, the toy appears to be balanced on an object, but in reality its center of gravity-the ends of the animal's balance pole, or the weighted tips of the bird's wings-is hanging below the object's base. These situations are fine demonstrations that will get students engaged in thinking about balance, center of gravity, and stability.

Materials

- Meter sticks: At least one per group. Be aware that the meter sticks will be abused in this activity.
- Clamps: One relatively heavy "C" clamp per group.
- Straight metal rods: At least one per group. Straightened wire hangers are perfect. Cut the ends off of the hanger, as they can catch in clothing and hair.

Teacher's guide

One Students should describe, in their own words, how they balanced the stick by keeping its base under its center of gravity.

Two Balance means to cancel the force of gravity with some equal force in an upward direction. This happens when your hand pushes up on the base of the meter stick, while the base is under its center of gravity. Tipping means the force of gravity is not canceled because the center of gravity is outside the meter stick's base causing the top end of the stick to rotate around it. If students are having difficulty with the idea of the stick rotating, as opposed to "falling," explain that the base of the meter stick is resting on their hands. When the stick becomes unbalanced, it does not just drop to the ground. Instead the top moves down, describing an arc of a circle and thus rotating around its base. If necessary, show this by guiding a meter stick through its "fall."

Three Horizontally. The meter stick is stable when laying horizontally—the base is the width of the hand and the center of gravity is very close to the hand. The

The High jump

You may want to end this activity by showing how high jumpers can clear a bar without their center of gravity ever going over the bar. In other words, their body went over but their center of gravity goes under! Humans can do this by distributing their mass (moving their bodies) so that their center of gravity is below their body like the bent metal rods in this activity.

meter stick is unstable when held vertically—the base is very small compared to the height of the center of gravity. Students should note that they prevented the rotation of the stick by placing the center of the stick on their palm, thus putting the center of gravity directly over the base.

Four The clamp gives the object much greater rotational mass. The stick will tip more slowly, therefore allowing students more time to move their hand to keep the base under the center of gravity.

Five Answers will vary, but students will probably predict the clamp at the bottom will be easier to balance.

Six Students should discover that the object is easier to balance when the clamp is up high. Perhaps they will even notice it rotates more slowly when the clamp is on the top of the meter stick. This is due to the object's large rotational mass. By moving the clamp to different points on the stick, students manipulated two variables: the object's center of gravity and its rotational mass.

Seven Students should discover that the balance point is a very small area but, once the point is located, the rod is easy to balance.

Eight The bent metal rod is quite stable and very easy to balance. Students should recognize that the bent rod is really hanging from their finger, not balancing above it. It is easier because the center of gravity of the bent rod is below the base or balance point.

Nine Answers will vary. Students should express that they balance an object by keeping its center of gravity over its base.

Extensions

Extension one Ask students why they put their arms out to help them balance.

Extension two Ask students why they think tightrope walkers carry long, flexible, massive poles.

Extension three Have your students hold a meter stick by placing an index finger under each end of the stick. Ask them to slide their fingers evenly towards the center. Only one finger will slide at a time, and which finger that is changes in a regular manner independent of their control. This is a fine challenge to explain. The less weighted finger will have less friction and will slide until the other finger becomes more weighted, at which time that finger will slide. It is important that students remember that the meter stick's center of gravity will remain relatively in the same spot.

Extension four If your students are interested in the high jump analysis, you may want to discuss these types of track-and-field activities to explore how people get their bodies over objects. The coaches and athletic trainers at your school may be a good resource for these investigations.

Extension five Have students choose one balanced toy, event, or balancing act, and explain how it works.

Activity 2

Balanced humans

How do you balance yourself?

Challenge

How does the human organism control its balance? What happens when we loose our balance? Why can balancing be difficult sometimes?

Objectives

- To describe how the human organism adjusts to maintain balance when standing.
- To describe how the human inner ear regulates balance.
- To describe the relationship between the human body's center of gravity and base.

Materials

- Timing device
- Carpenter's level
- Human biology resources

Procedure

One How stable are you? Explore your stability, and that of the other students in your group, by standing in the following positions, with your eyes both open and closed.

- Feet Apart: Stand straight with your feet at least shoulder width apart.
- Feet Together: Stand straight with your feet together.
- Foot Up: Stand on one foot, and raise your other foot by bending your knee so that your raised foot is at least as high as your knee.
- Step, Crouch: Move one foot a step forward and crouch down.
- Straight Leg Up: Stand with one leg raised directly in front of you.
- Straight Leg Side: Stand with one leg raised to your side.

Create a table. Record the time for each trial, with eyes open (O) and closed (C), and rate each position as either stable (S) or unstable (U). Determine the horizontal and vertical position of each group member's head using the carpenter's level, and record that information separately.

Two Describe the difference between balancing with your eyes open and closed. Speculate on the causes of these differences, and record your thoughts.

Three Which situation is the most stable? Which is the least? Why?

Four What did your body do to maintain its balance?

Five How was your head positioned in each situation?

Six Did you have to consciously think about keeping your balance? Explain.

Seven How do you balance yourself?

■ TEACHER-TO-TEACHER

This is a realistic and fun activity. Students explore their own physical stability in several positions, describe the body's center of gravity-to-base relationship, and relate their observations to how their inner ear controls balance. Your students' prior knowledge of the human inner ear, their level of interest, and the resources you have available, should determine how deeply you are able to discuss the human biology involved in balance. It will take one to two periods to perform this activity.

Materials

- Timing devices: Student watches or the classroom clock. The relative stability of each position, and not accuracy of the times, is important.
- Carpenter's levels: Several short (~15 cm) levels will work well.
- Resources for students to research how the human organism controls balance: A copy of "Balance and the Human Ear" (page 55) may be all that is necessary.

Teacher's guide

One Results will vary (see Figure 2.1). You may choose to set a maximum time limit, perhaps one minute, as some students may be able to balance in these positions for a while. You may also wish

GROUP MEMBER	EYES	FEET APART	FEET TOGETHER	ONE FOOT UP	ONE STEP, CROUCH	STRAIGHT LEG UP	STRAIGHT LEG SIDE
Name	Open	30 sec.	30 sec.	~30 sec.	30 sec.	~30 sec.	~30 sec.
	Closed	30 sec.	30 sec.	~15 sec.	30 sec.	~10 sec.	~5 sec.
Name	Open	30 sec.	30 sec.	~30 sec.	30 sec.	~30 sec.	~30 sec.
	Closed	30 sec.	30 sec.	~15 sec.	30 sec.	~10 sec.	~5 sec.
Name	Open	30 sec.	30 sec.	~30 sec.	30 sec.	~30 sec.	~30 sec.
	Closed	30 sec.	30 sec.	~15 sec.	30 sec.	~10 sec.	~5 sec.
Name	Open	30 sec.	30 sec.	~30 sec.	30 sec.	~30 sec.	~30 sec.
	Closed	30 sec.	30 sec.	~15 sec.	30 sec.	~10 sec.	~5 sec.
STABILITY	Open	Stable	Stable	Unstable	Stable	Unstable	Unstable
	Closed	Stable	Stable	Unstable	Stable	Unstable	Unstable

FIGURE 2.1
Sample student data table.

to have the groups develop rules for the use of their arms while balancing. If so, they should note their rules and reasons in their activity write-up. Explain that the carpenter's levels should be used to record whether their heads tilt while they are balanced. Data such as "head tilted forward, chin down" is sufficient. Students do not need to calculate averages, but should use the data from the entire group, not just themselves.

Two Generally, at first, it is harder to maintain balance with one's eyes closed. Students should infer that sight plays an important part in our ability to balance. Some students may mention that blind, or otherwise visually impaired people, probably have a "normal" sense of balance and do not fall over when they try to balance. They may then conclude that with time and practice, they could learn to be more stable with their eyes closed. (See Extensions one and two.)

Three Responses will vary, but the crouched stance should be cited as the most stable because it has a large base and low center of gravity. Students should cite any of the one-legged stances as being least stable because they each have a small base and high center of gravity.

Four Answers will vary, especially if you had group's design rules to govern moving their arms to maintain their balance. Student answers should demonstrate they understand the need to keep their center of gravity over their base.

Five In most cases, people tend to keep their heads straight up and horizontal while balancing. The carpenter's level can give students data to refer to when they explore how the inner ear functions.

Six Generally people do not have to think about balance, and their bodies automatically adjust to changes in position. However, by closing their eyes, students were probably more aware of their bodies and will say they thought more about balance.

Seven Responses will vary with your students' knowledge, interest, and available resources. You should ask them to use their own thoughts and experience in their answers. This particular question could be used as an independent project, or as the basis for one. (See Extension four.)

Extensions

Extension one The development of the human ability to balance is very interesting, and could be used as a topic of discussion or take-home assignment. Some students may have infant siblings, or other opportunities to observe infants and toddlers either crawling or learning to walk. You might ask how an infant's body size and proportion affect success in learning to walk. You may also want to discuss physical ailments unrelated to the ear that cause walking problems. Be aware that this extension may cause embarrassment for any student who is "awkward" or clumsy, or has a restrictive mobility handicap.

Extension two It is unlikely that a person will forget how to walk. However, we learn to balance over and over again in our lives—riding a bike, rollerblading, skateboarding, skiing, snowboarding, and surfing. Do your students think that age affects one's ability to balance, or to learn a new balance-related activity? Why?

Extension three Imax movies provide an excellent example of how human sight plays a part in human balance. Although the audience is sitting and the seats are immobile, the combination of a special camera and a nearly circular screen makes audience members feel as if they are moving. Some audience members experience actual motion sickness. However, if they close their eyes, the feeling goes away.

Extension four You might choose to focus a report or project on the inner ear. Topics could include: explaining why we get dizzy; why some people feel ill when they become dizzy; how our health—from the common cold, or ear infection, to disease—affects our balance; and how physical occurrences, like a perforated ear drum, might affect our balance. Caution students to be careful if they choose to experiment with their own dizziness. (See "Do You Hear What I Hear" (page 101), and "Testing: One, Two, Three" (page 117) for more information about the human ear.)

Activity 3

Balanced vehicles

Why don't vehicles tip over on steeply banked turns?

Challenge

Why are some roadway turns banked? Why do turns on banked roads feel smoother than turns on flat roads? Why do motorcycle riders, automobile passengers, and other people in motion lean into turns?

Objectives

- To describe the forces acting on a vehicle and its passengers during a banked turn.
- To explain how banking or leaning into a turn results in a balanced situation.

Materials

- Toy truck
- Plastic container
- Materials for lining the bottom of the container
- Loads for the back of the toy truck
- Wooden rod with capped end
- Protractor

Procedure

One As a group, collect your materials. Draw diagrams of both an end view and a side view of your toy truck.

Two Devise and record a method for measuring the center of gravity of your toy truck. Measure the center of gravity of your toy truck and mark it on each of your drawings.

Three Find the maximum angle your toy truck can lean to one side without tipping over. Use a protractor to measure that angle, and draw a force analysis diagram of an end view of the angled toy truck. Label the angle.

Four Vehicles often carry loads that significantly change their center of gravity, thus affecting their tipping angle. Using the loads provided, re-examine the maximum angle your toy truck can lean without tipping over or losing any of the load. Measure those angles and make a force analysis diagram of an end view of the toy truck with at least two different loads.

Five Place your toy truck in the center of the clear plastic container. Raise one side of the container, and measure and record the angle at which the truck begins to slide. Repeat this experiment several more times, each time with a different material between the truck and the bottom of the container. Describe the difference in behavior of the truck for each trial. Why does placing different materials under the truck result in different responses by the truck? Why are these situations more like the real road than when the truck is on the plastic?

Six You have been examining the effects of tipping and banked roadways while your toy truck is stationary. In order to explore a more life-like example of a vehicle traveling through banked turns, use the wooden rods provided to spin the container with the truck inside it. (See Figure 2.2.) What is the effect of a banked roadway on the tipping of your toy truck while going around a turn? **Exercise caution while spinning the container.**

Seven Draw a force analysis diagram of your toy truck turning on a banked roadway.

Eight Complete a force analysis diagram on the sketch of the cyclist at the beginning of this activity. Why do cyclists' lean into turns?

Nine Why don't vehicles tip over on steeply banked turns?

No friction? No problem.

In order for a vehicle to turn, it relies on the friction created between its tires and the road. But what happens when the amount of friction is greatly reduced? This could result from wet or slippery roads, or from tires that are not spinning because the brakes have locked. The Anti-lock Braking System (ABS) was created to aid drivers who had to stop quickly. Before ABS, drivers would have to pump on their brakes to prevent the brakes from locking, even though the normal response to a potential collision is for the the driver to "slam" on the brakes. With ABS, the driver could do this and the ABS would pump the brakes for him/her. In 1998, ASHA Corporation announced a new traction technology called Gerodisc. Gerodisc limits wheel spin if traction is reduced by sensing changes in wheel speeds and, in response, is able to shift power to the non-spinning wheel in order to regain traction.

What other variables affect vehicle stability in turns? What parts of an automobile would you modify to increase its ability to maintain traction on the road?

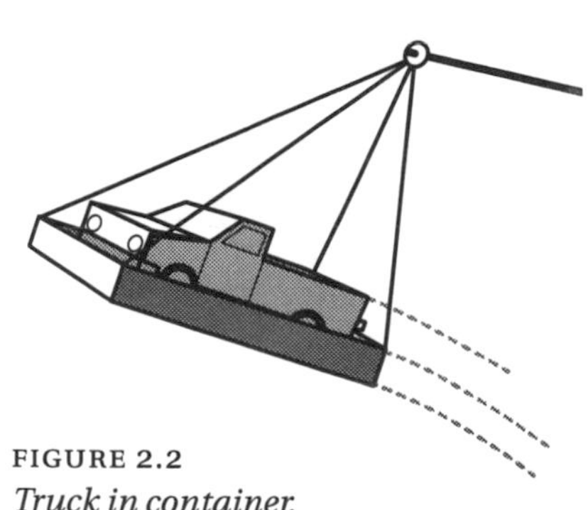

FIGURE 2.2
Truck in container.

TEACHER-TO-TEACHER

This activity has students simulate moving on a banked roadway by spinning a toy vehicle in a container. In order to introduce the forces involved, which are very difficult to describe intuitively, find out if any of your students can relate personal experiences to this activity. For example, someone may have been in a vehicle that skidded through a turn on a dry road, or was unable to make a turn and was forced off the road. (See Extension three.)

This activity will take three class periods. You will probably need to do some direct teaching of the force analysis while a vehicle moves on a banked roadway. It is easy to get caught up in the details of the force analysis. Try to make it a big picture analysis. The process of using diagrams and asking and answering questions about the forces in conceptual terms is much more important than the actual force analysis.

Materials

- Plastic container: One container, approximately 20 cm × 15 cm × 9 cm, rectangular, with a clear, flat bottom, per group. Clear plastic 64 fl. oz. food containers from the deli section of a grocery store are perfect. An opened paper clip with an end heated-up can be used easily to make the holes for the string. Make two holes per corner, placing one on the top flap and one on the side wall for a more secure connection.
- String: Two 1m pieces of string for each container. Tie one free end to each corner, then use an overhand knot to tie the centers of the two strings into a loop about 40 cm from the container. The loop's diameter should be small enough to keep the string from sliding off the rod while the container is being rotated.
- Toy truck: Plastic truck about 17 cm long, or small enough to fit into container.
- Materials to cover the bottom of the plastic container: Pieces of sand paper of various grades cut to cover the entire bottom of the container are perfect, and the smooth back of the sandpaper can be used as well. You may also wish to involve your students in the process of choosing these materials, in order to further explore the concept of friction, and how different degrees of road "roughness" and surface area affect friction between the tire and the road.
- Loads for the back of the toy truck: Small balance weights work well, as do sugar cubes. Objects must be small enough that several of varying mass can be placed in the truck.
- Wooden rod with end cap: One 40 cm long, 1 cm wooden dowel, with a 2.5 cm fender washer screwed into an end for each group.
- Protractor: One for each group.

Teacher's Guide

One Diagrams will vary. Figures 2.3 and 2.4 are examples. "×" marks center of gravity.

Two Methods will vary. Students may try balancing the truck on the edge of a book, on their hand, or on the tip of their pencil, and several may draw a connection between the balance point and the center of gravity. Some students may determine that the balance point is the center of gravity. It is important that students discover that, because the front of the truck is heavier than the back (depending upon the truck), the center of the truck is not necessarily the center of gravity. (See Figure2.3.)

Three Diagrams will vary (see Figure 2.5). Ensure students correctly label the angle and forces. Students should find that, at the tipping point, the center of gravity should be directly over the center of the tires still on the ground. If students have difficulty finding this angle, have them place the truck on the table and begin to lift one side up, with their other hand on the other side of the truck. They should measure the angle at which the truck is almost about to tip over, which can be found by tipping it into their other hand a few times.

Four Force analysis diagrams will vary. Students should indicate that the center of gravity and angle of lean will vary according to the load in the truck.

Five Students will discover that the additional friction provided by materials with rougher surfaces will result in a greater angle before the toy truck begins to slide. They may find the friction to be great enough to get the toy truck to a tipping angle before it slides. With this step, students should definitely see the relationship between the surface texture of the road, friction, and the ability of a vehicle to stay on the road.

Six Students should discover that the truck does not slide or tip over when rotating around the wooden rod as diagrammed below. You may need to demonstrate how to get the toy truck/plastic container safely rotating around the rod. The banked roadway tilts the toy truck into a lean, thereby helping to move the toy truck through its turn.

Seven Force analysis diagrams will vary. (See Figure 2.6).

Eight Students should include the force of the cyclist's lean in their force analysis diagram. If students have difficulty addressing this, encourage them to consider the cyclist and bicycle as a single object, and to think in terms of center of gravity and base. When the cyclist turns the bicycle, he/she is moving the base out from under the center of gravity. At the same time, the object is still moving forward. By not leaning, and matching the motion of the top of the object with the base, the object will tip over-the base, wheels of the bicycle, will essentially accelerate out from under the rest of the object, the cyclist. If necessary, revisit "A Balancing Act" (see page 35), and remind students how they tried to keep the base of the meter stick beneath

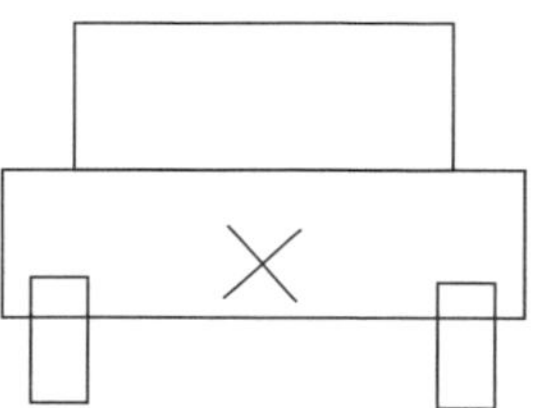

FIGURE 2.3
End view.

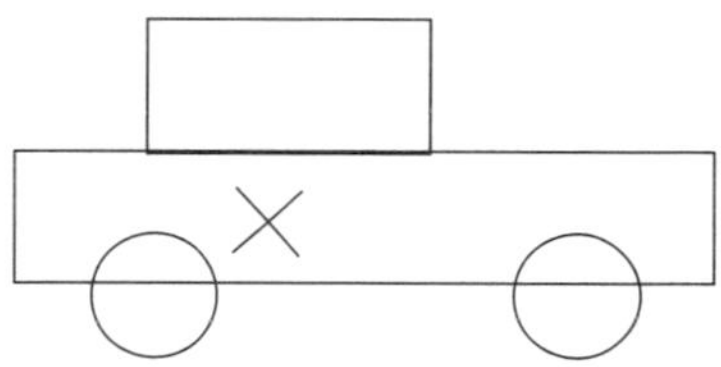

FIGURE 2.4
Side view.

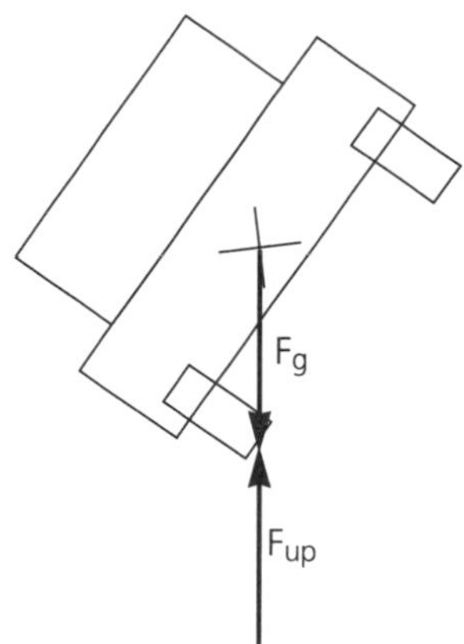

FIGURE 2.5
Angle of lean.

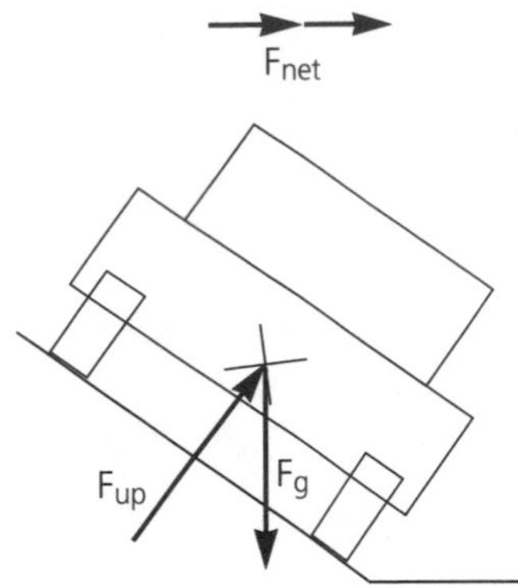

FIGURE 2.6
Turn on banked roadway.

the top. A second concept involved in this scenario is that of the force of the lean. Because the cyclist leans, or tips, into the direction of the turn, the force of the lean helps the turn. (See "Forces that Turn You" page 56.)

Nine Students should understand that, in a banked turn, the road physically tips whatever is traveling on it into a lean. (See Extension two.)

Extensions

Extension one Extend this activity into a field project by actually measuring the angle of local roadways and arcs of turns. Students will find that different turns are posted with different speeds, which relate to the arc of the turn and the degree of banking. You may be able to contact civil engineer at the local Department of Public Works, who might assist you and your students.

Extension two Plan a trip to a local auto or bicycle racing track, if one is available. Students will enjoy seeing just how steep some banks are, and how fast an automobile or bicycle needs to be traveling to use the banking effectively.

Extension three Use scenes from popular movies, or other video materials (see Resources, page 146), to demonstrate cars skidding along in their initial direction instead of turning smoothly. You can diagram these situations using circles and tangent lines to show students what is happening.

Activity 4

Describing balance in turns

How is balance different when turning?

Challenge

Do vehicles in motion remain stable? What is the relationship between the physical sensations you might experience when turning, the forces acting on you, and the stability of your vehicle? Can you explain the stability of a vehicle to your peers?

Objectives

- To use a force analysis diagram to define forces acting on an object while it is both moving forward and turning.
- To relate the forces acting on a vehicle to the stability of that vehicle.
- To connect physical sensations experienced while turning to the forces that cause them.
- To analyze a stable turning vehicle.

Materials

- Stability Analysis Packet
- Clear transparency sheet and pen
- Large, paper arrows to represent forces

Procedure

One As a group, review the scenario your teacher assigns to you. Describe the stability of the vehicle in your journal—do not write on the group summary sheet.

Two On your individual stability analysis sheet, construct a force analysis of the situation by sketching the forces you think are acting on the vehicle. Consider the relationship between the forces and the base of the vehicle. Record your ideas in your journal, and answer the questions on your personal stability analysis sheet.

Three Share your individual stability analysis sheets with your group. As a group, discuss the merits of each person's ideas.

Four Start to develop a stability analysis, in pencil, on your group summary sheet.

Five Ask your instructor to review your stability analysis.

Six Finish writing out the stability analysis on your group summary sheet, making any necessary adjustments.

Seven Trace your group's scenario stability analysis onto a clear transparency using the special transparency pens provided.

Eight As a group, prepare a presentation which describes the stability of the vehicles in your group's situation in terms of forces acting on the vehicle with respect to the vehicle's base. Be certain all group members are involved.

Nine Present your stability analysis to the class. Use the large arrows provided to represent each force in your presentation.

■ INDIVIDUAL GROUP PACKET

Instructions

One Review the scenario and come up with a group stability analysis.

Two Draw and label the forces acting in the scenario.

Three Using dashed lines, indicate the horizontal and vertical components of these forces using the road surface as horizontal.

Four Mark the base of the vehicle.

Five Describe the stability of the vehicle.

Six Describe the sensations you would feel in these situation in terms of the forces you've described.

Presentation outline

Each group member should participate and have a role in the presentation.

I Each group member introduces themselves

II Describe the situation and indicate the force of gravity

III Describe and indicate other forces acting in the situation

IV Describe the net force and stability of the vehicles

V Connect the human sensations involved to the forces

VI Questions and answer time

■ INDIVIDUAL STABILITY ANALYSIS SHEET

Procedure

One Sketch the forces you think are acting on the vehicle in the scenario. What is the net force? Is the vehicle stable? Why?

Two Reflect upon your individual analysis and your group's final analysis.

Three Have you ever experienced the motion illustrated in any of these situations? Describe the physical sensations you felt in the situations you have experienced.

Four What does stability mean to you?

■ TEACHER-TO-TEACHER

This applied activity and presentation may be difficult for some students to do. Depending upon the students in your class, you may want to demonstrate a stability analysis for them. For the turning situations, a fine student explanation might be something like "the force of gravity adds to the force upward from the roadway and the friction force to result in the force that turns a vehicle." The stability issue should be easy for the constant speed situation and appear to be tipping or falling in turning situations. Students should be able to say "the vehicle is falling into the turn" in a somewhat convincing manner.

To assemble the Group and Individual Analysis Packets, you will need to photocopy pages 60–64. The Individual Packet should contain a copy of the scenario and a copy of the appropriate procedures and questions (page 52). Students should record responses on packet pages, as well as in their journal. Student preparation will take one period, and presentations will take one to two periods.

Materials

- Stability Analysis Packets: One per group.
- Clear transparency sheets and colored pens for writing on transparency paper
- Overhead projector
- Arrows: Five to ten, 10 cm wide by 30 cm long, for students to use in their presentation.
- Meter stick or pointer for use in presentation

Teacher's Guide

One Check with each group to ensure they understand the diagrammed scenarios.

Two Responses will vary and are not intended for grading. Remind students to record their thoughts in their journals.

Three Keep these conversations positive!

Four Force analyses will vary with each situation, but arrows should be accurately drawn.

Five Be certain to check off each group analysis. Correct, or make suggestions, as necessary.

Six Ensure that each group has a good analysis before they start planning their presentation.

Seven Distribute transparencies and pens as you approve each group's final analysis.

Eight (The Presentation Outline included in the Packets is intended as an example, and may be modified to your own specifications.) Responses will vary. Be certain to remind groups that all members should participate in creating the presentation. If necessary, you may want to choose those students who will make the presentations.

Nine Be sure to provide time for groups to practice. These presentations should be a successful experience. Supply a pointer and large arrows for each group as they make their presentation, and encourage the rest of the class to ask questions.

Extensions

Extension one Some of your students may have seen yellow caution signs that warn of turns and suggest slower speeds. Some may have seen signs that indicate that a truck may tip over on a turn. Have your students do a force analysis of a large truck—18 wheeler or larger—traveling forward and then making a turn. Encourage them to compare this type of truck to the pick-up type they used in "Balanced Vehicles" (page 44). Remind them of the differences in stability they discovered in "Balancing Act" (page 35), and have them include force arrows indicating rotational mass of a truck that is empty, half full, and packed with heavy objects close to the top. Why might a truck tip over in a turn? Considering rotational mass, which of the three above situations would be more likely to tip over? Which would be the least? Why? Is a car as likely to tip over as a truck? Would a banked turn make a difference?

Extension two Have students take notes on their peers' presentation, and write short assessments.

Background reading

Balance and the human ear

Our body's sensory system works very hard to maintain our balance, or equilibrium, and has many sensory units that give us information about our body's position and movement. This information gives us the conscious and unconscious ability to maintain our balance.

Throughout our bodies are **proprioceptors**, sensory receptors that tell us where our body is and whether it is changing motion. These sensory units are located in and around our joints, muscles, and skin, and in our inner ear. The many sensors located throughout our body give us information that allows us to react, often by reflex, in order to maintain our balance. When we are standing still, the pressure sensors in our feet signal slight pressure differences that we react to automatically. These sensors directly stimulate appropriate muscles to keep our center of gravity over our base, and tell our brain what is happening. We can "think" about these signals, or simply ignore them and let our bodies work. Automotive technicians are copying this natural system by regularly adding sensors to vehicles which will enable them to react appropriately and/or signal the driver when the vehicle becomes unbalanced. The ability for an organism to know where it is and in what state its parts are in is called **proprioception**.

While the sensors throughout our body help maintain our balance, our inner ear regulates balance by sensing the motion and the position of the head. The two parts of the inner ear responsible for this regulatory function are the **semicircular canals** and the **vestibule**.

The semicircular canals are composed of three loops or canals filled with fluid at right angles to each other, much like a carpenter's level with three axes instead of the customary two. In a carpenter's level, a bubble in the fluid indicates when the level is perpendicular to the force of gravity, thus helping determine when an object is horizontal or vertical. The semicircular canals contain fluid that moves as our head accelerates either front-back, left-right, up-down, or any combination of these motions. At the base of each canal are **ampullae**, bulb-like structures, which contain **cilia**, hair-like structures. The motion of the fluid bends the cilia, causing nerve impulses that our brain interprets as acceleration. This information helps us keep our heads straight up and level.

The vestibule is located between the semicircular canals and the **cochlea**—the section in which sound vibrations are converted into nerve impulses. The pouches in the vestibule have cilia that send nerve impulses to our brain like the cilia of the semicircular canals. However, these pouches are filled with a jelly-like substance upon which float granules of calcium carbonate called **otoliths**. When the otoliths move, they cause the jelly-like substance to move, which bends the cilia, and signals our brain that our head has accelerated. For example, when we tilt our heads forward, the force of gravity pulls the otoliths downward, in turn dragging the jelly-like substance which then bends the cilia. The same effect occurs when we accelerate horizontally. The base of the vestibule accelerates out from under the otoliths dragging the jelly-like substance resulting in the bending of the cilia. When the cilia bend, nerve impulses are sent to our brain and are interpreted as acceleration. In respond to this, humans tend to lean forward, perhaps because our brain anticipates motion and wants to prevent our feet from accelerating out from under us.

To maintain our balance we need a great amount of information and control. Our body manages this complicated task with a combination of all of these sensory units. The most important, however, is probably the inner ear. Motion sickness is a disturbance of the inner ear's balance-sensing areas.

Forces that Turn You

So you are driving along in a car with your dusty science book on the dashboard. You turn right and the book slides to the left. You turn left and the book slides to the right. Was it pushed by some force, or did the car slide out from under the book?

Before describing the forces that turn you in a car, consider the forces that stop you and the car. If you were in a car, and a friend was watching you from the street, and the car stopped suddenly, you might both agree that friction forces stopped the car. But what stops you? If you were not firmly attached to the car with a safety belt, you'd be thrown forward. From your perspective, you were sitting still one moment and thrown forward the next. You might say that some force pushed you. Your friend, however, would say that when the car stopped moving, you didn't. If this person were correct, then the force you described could not be real.

A similar situation occurs when you turn. For example, you know that forces are involved in turning both your car and you to the right. You feel the forces in the turn, and might conclude the following:

"While turning right a force pushed me to the left. This force was directed out from the center of the arc of the turn, so I'll call it a center-fleeing or **centrifugal force**."

However, your friend might say the following.

"The car pushed you to the right, guiding you through the turn, otherwise you would have continued moving forward. There is no force to the left. Because the force of the car against you acts toward the center of the arc of the turn, I'll call it a center-seeking or **centripetal force**."

So, who is right? Are you both right because it is just an action-reaction thing? Well, as you learned in "Notions of Motion," it can not be an action-reaction thing because both these forces are acting upon you and would therefore cancel! (Remember action-reaction force pairs always act on separate objects.)

The physics you know does not work well when you are accelerating, because our normal state is not one of acceleration. Instead, we make up forces to explain the motions we see/feel while accelerating. You made up a force to explain the dusty science book sliding across the car's dashboard, and that force is just that, made up. By its definition, centrifugal force can only exist during acceleration. During acceleration, however, centrifugal force does not really happen, and therefore does not exist.

Let's describe the forces that turn you with a force analysis diagram. Consider a force analysis diagram (Figure 2.7) of a car that is not turning. Notice that the sum of two forces—two because we are analyzing the end view—up (F_{up}) from the ground cancel the force of gravity (F_g) downward. The net or sum of all forces (F_{net}) acting on the car is zero, so the car is not speeding up, slowing down, or turning.

Now let's look at a car turning to the right (Figure 2.8). Notice that the friction force (F_f) acts on all tires but is just diagrammed as a force acting on one tire. This is an example of the force analysis diagrams in Activity 4 of "Stability when Turning" (page 49).

There are many ways you could describe the net force acting on this car. It is the sum of all forces or the net force, it's caused by the interaction between the tires and the road so it is a frictional force, and it causes a circular-type motion so it is a centripetal force. If the frictional forces are not great enough to turn the car, the car would travel straight along a line tangential to the arc. The car would not go directly out along a radius line, which is what a centrifugal force would cause. You might want to explore this motion in a safe laboratory setting before you experience it in a car trying to turn too quickly on a slippery road.

Also notice that the turning force—the friction on the tires—does not act on the car's center of gravity. Thus, the tires tend to accelerate or turn under the car, and the car tips. If the tipping is great enough to cause the center of gravity to move outside the car's base, it becomes unstable and tips over, as in "Balanced Vehicles" (page 44).

In Figure 2.6, you can see that a banked roadway helps turn a car in two ways. First, the banked road actually pushes a bit in the direction of the turn, eliminating the complete reliance of the vehicle upon the frictional forces between the tires and the road. Second, the tilting of

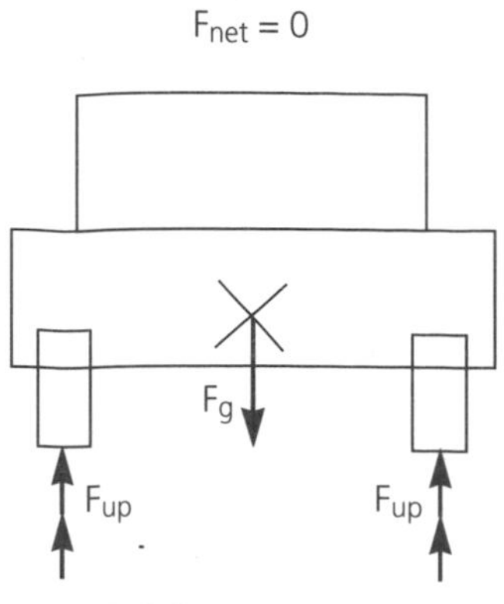

FIGURE 2.7
v = constant
Car is stationary or is moving at a constant speed in a straight line.

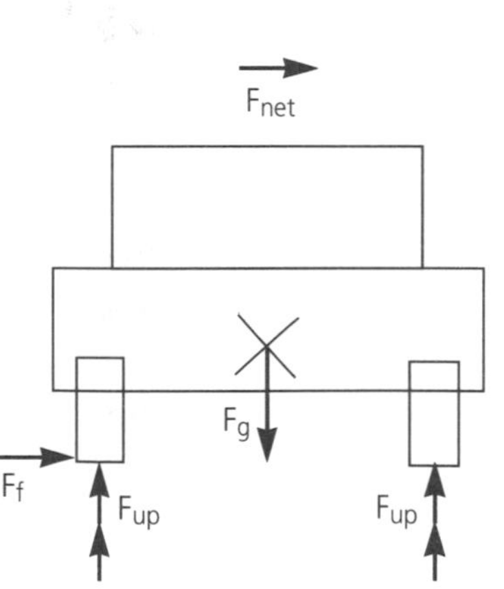

FIGURE 2.8
Turning.

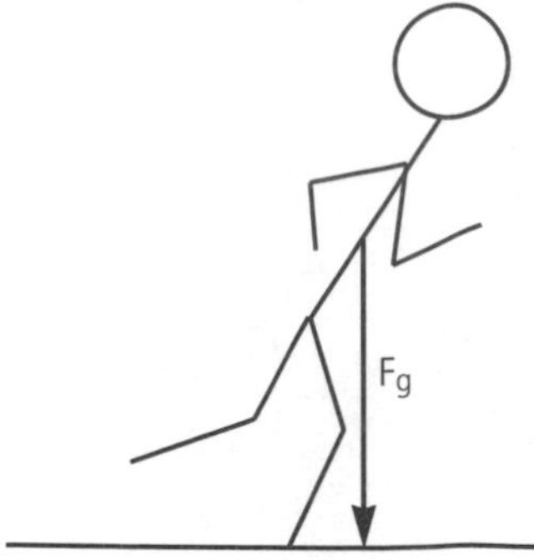

FIGURE 2.9
Side view.

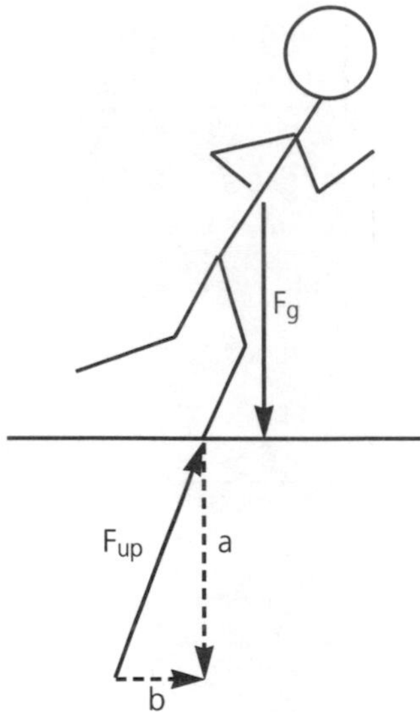

FIGURE 2.10
Turning on flat ground.

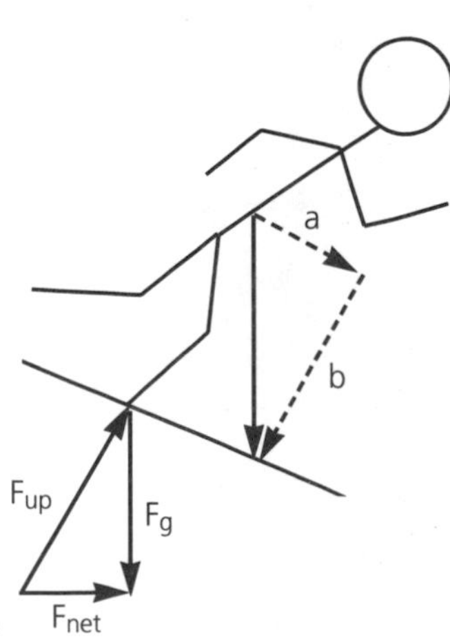

FIGURE 2.11
Turning on banked ground.

the car puts its center of gravity more over the inside tires, reducing the risk of tipping over.

Leaning Into a Turn

Force analysis diagrams do not explain why you do not tip over when leaning into a turn.

In this side view of a runner turning left (Figure 2.9), the center of gravity is clearly outside the base, yet the runner does not tip over. Why might that be?

Before considering the turn as diagrammed above, consider the start of a race. You lean forward so much so that your center of gravity is no longer over your base, and you have to put your hands down or else you will fall forward. When the race begins, you accelerate your feet under your body while you are falling resulting in a somewhat neutral situation. If your feet slip, you fall forward. If you did not lean forward, you might accelerate your feet out from under your center of gravity and fall backward.

Thus the forces acting on you while you are moving do not all act directly upon your center of gravity. Different parts of your body are moving at different rates. To keep your center of gravity above your base, and to maintain your balance, sometimes you must lean.

For example, picture a person running through a turn. Their feet accelerate through the turn, but their body continues forward. This creates a situation in which the body's center of gravity may move outside its base. Without compensating, the runner will tip outward into an unstable position. In reality, the runner has learned to lean into a turn to maintain stability, positioning his/her body to allow his/her feet to accelerate under his/her center of gravity, thus maintaining a neutral balance situation. If the acceleration is not great enough, the runner has overcompensated and tips, or falls inward. Consider the force analysis diagrams (Figures 2.10 and 2.11) of a runner turning to the left on flat and banked ground .

On the flat ground, the vertical component (a) of F_{up} cancels the runner's F_g while the horizontal component (b) is entirely caused by friction between the feet and ground resulting in the net force turning the runner. If the friction is not great enough, the runner's feet slip out from under him/her.

On a banked turn the F_{up} is quite close to a force normal (90° to the ground) and supplies the necessary turning force. However, consider the components of the runner's force of gravity. The normal component (a) of F_g is canceled by F_{up} and the component parallel to the ground (b) must now be canceled by friction. Banked turns are designed so that less friction is needed compared to flat turns.

Slipping occurs when there is not enough friction between the ground and your feet; for example, imagine trying to run across a smooth floor wearing socks. On a flat surface, friction provides all the force to speed you up, slow you down, and turn you. Banking a turn can greatly reduce the risk of slipping when turning. Similarly, you can eliminate the risk of slipping at the start of a sprint by using starting blocks that allow you to push off more in the direction you wish to go, again reducing the necessary friction.

In summary, we lean into a turn so that our base—feet—does not turn under our center of gravity—torso—causing us to tip outward. The resulting situation is a neutral balance while turning. Banking a turn lets us push off more in the direction we wish to turn and rely less on friction.

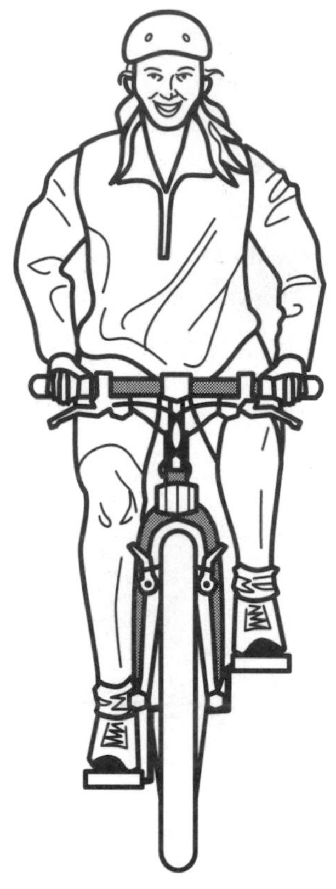

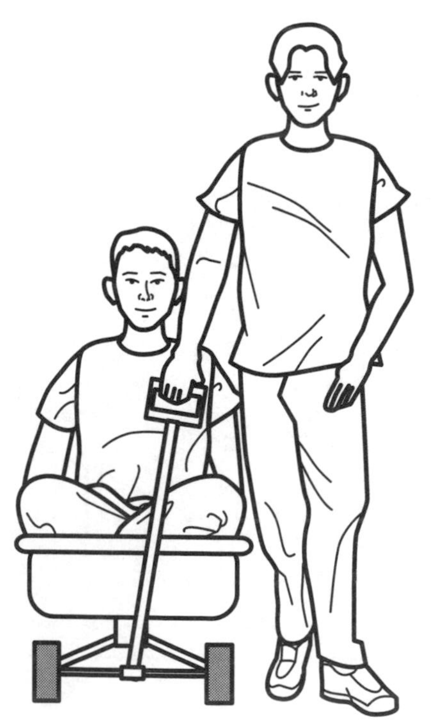

chapter three

Energy and Reaction

■ TEACHER-TO-TEACHER

In this learning plan, students will review individual concepts they may have learned in other classes and combine them in new ways. They will be asked to relate food calories to mechanical energy, and to think of the human body as a machine. Students will be challenged to think of the energy involved in human motion and compare it to the kinetic energy of a vehicle. They will perform energy analyses of an in-class system, which includes human and mechanical components and is analogous to "real life" transportation situations. Students will use them to look at some larger transportation situations in terms of energy, and communicate their learning to their peers.

Conceptual development

Food provides the human organism with energy that is transformed into life functions, thinking, and mechanical work. Food energy is measured as available heat in Calories, whereas mechanical energy is calculated as work in Joules, and measured as force acting over a distance. Energy transfers are never perfect or 100 percent efficient. But, the efficiency of the many energy transfers involved in riding a bicycle can be used to determine how much work can be done from one food item. The following paragraphs describe the concepts developed in the four activities of this chapter.

Kinetic energy is directly related to a vehicle's mass and, more importantly, exponentially related to its speed—twice the speed is four times the kinetic energy. The work done to start and stop a vehicle is directly related to its kinetic energy, therefore the factors of this work are directly related to a vehicle's kinetic energy.

Conceptual construction

Prior experiences of human and vehicular fuel combined with knowledge of energy transfers.

Central Idea Energy is transferred in many ways.

- **Food and Fuel** Explore human and vehicular energy sources.
- **Energy of Motion** Investigate a vehicle's kinetic energy changes.
- **Energy Transfers** Investigate a vehicle's energy transfers.
- **Speeding Up and Stopping** In teams, analyze transportation energy transfers.

Constructed and Communicated Idea Transportation systems have many energy transfers.

The efficiency of the human organism is very low considering the mechanical work it can do—just one Calorie of food is a lot of mechanical energy. Energy can be transformed and these transformations can be analyzed. Energy analyses can illuminate relationships that are obscured in the details of force and motion analysis.

Using an analogous situation, the effect of manipulating energy-related variables can be investigated. Simple laboratory activities can illuminate the ideas involved in complicated transportation situations.

Theoretical science can be applied to everyday transportation situations.

National Science Education Standards

In the development of the conceptual construction of this chapter, the following Content Standard concepts and principles served as a guide.

- Energy stored in bonds between the atoms (chemical energy) can be used as sources of energy for life processes (page 186).
- Energy is transferred in many ways, however, it can never be destroyed (page 180).
- All energy can be considered kinetic, potential, or energy contained in a field (page 180).
- Energy can be transformed by collisions in chemical and nuclear reactions, by light waves and other radiations, and in many other ways (page 180).
- Design and conduct scientific investigations. Formulate and revise scientific explanations and models using logic and evidence (page 175).
- Communicate and defend a scientific argument. Understanding about scientific inquiry (page 176).

Time management

This learning plan explores important ideas, and moves from qualitative understanding to quantitative analyses. The readiness of your students will greatly affect the class time involved. For students with good mathematics skill and understanding, the entire learning plan may take 10 to 12 fifty-minute periods. Activity 1 is a group of three activities which may take one to two periods each, and should be concluded and reviewed in a separate period. Activity 2 will probably require some direct instruction and in-class discussion, and can take one to two periods. Activity 3 may take four periods, and Activity 4 will take three to four periods.

Learning plan assessment

The written products and presentations in this learning plan provide easily graded individual and group work. You could develop a point system, depending upon how you implement the activities, and students can be given credit as they perform each activity. Final grades would be the accumulated credits. The final activity of this learning plan provides a fine opportunity to implement performance assessment that involves students applying what they have learned in a situation they have been experiencing. Students can also perform peer reviews on each presentation. Create an assessment rubric (see Appendix, figure A-1) before the first presentation.

Activity 1

Food and fuel

How far can you ride a bicycle using the energy available in one food item?

Challenge

How does energy flow through your body? How are the Calories you eat related to the work you do? How much work can you do from just one Calorie of food? Does your body use energy to think about doing something and then to signal muscles to move? Are machines, like bicycles, more efficient than the human body?

Objectives

- To recognize that energy available in food is associated with heat measured in Calories.
- To discover the efficiency of the human organism in terms of energy input versus output.
- To describe the equivalence of Calories and Joules in a scientific manner.
- To discover the efficiency of a bicycle in terms of work applied and distance traveled.

Your group will explore one of the following investigations: food-to-work, human reaction, the bicycle. Follow the procedures below for your particular exploration. After your exploration, you will share your discoveries with the rest of the class in order to establish an energy picture of a person riding a bicycle.

Alternative fuels

Do you know where gasoline comes from? Although science has developed very efficient means for producing gasoline from oil, scientists are well aware that oil supplies will not last forever.

Scientists are working on many fuel alternatives, to both increase efficiency and reduce pollution emissions. For example, E85—15 percent gasoline and 85 percent ethanol—vehicles have been available to the general public since 1992, and millions are on our roads today. In fact, ethanol-blended fuel sales represent 11 percent of all automotive fuels sold in the United States. E85 is a high quality octane booster, and is capable of reducing emissions. Other alternative fuels such as dimethyl ether, compressed natural gas, and propane—currently the most widely used—are being studied as well.

If you were a scientist, what alternatives to gasoline might you research? What properties of these fuels would you be most interested in? Would need to design a new engine to accommodate a new fuel? What environmental ramifications would you have to consider?

FOOD-TO-WORK

Materials

- Food wrapper with nutritional information on it

Procedure

One What is the energy, in Calories, available in your food item (E_{in}). Staple the food wrapper into your journal.

Two What will your body do with this food that will enable you to pedal a bicycle?

Three How efficient is the average human being, in terms of energy consumed and work done? Record you reasons.

Four What is efficiency? The standard level of efficiency from one trophic level to the next is 10 percent—regardless of the amount of energy consumed at any level, only 10 percent can be passed on. Using this value for human efficiency, calculate the theoretical amount of Calories (E_{out}) your food item provides that are available to do mechanical work.

Five Mechanical work is measured in Joules, not Calories. How much energy, in Joules, is available in your food item (4,180 J = 1 Calorie).

HUMAN REACTION

Materials

- Stop watch

Procedure

One Your teacher will assign one student to hold the stop watch, and one student to record results.

Two Form a circle, with everyone involved in this activity holding hands and facing out.

Three The student holding the stopwatch should hold it in their right hand. The student to their right should hold the stopwatch holder's right wrist with their left hand. The stopwatch holder will start this investigation by clicking the stopwatch on. Once the student to their right feels this motion, they will squeeze the hand of the student on their right, and so on around the circle. The stopwatch holder should stop the watch when their left hand is squeezed. Repeat several times.

Four The recorder should record the number of people in the circle, and the time for each trial. If the recorded times vary a lot, the recorder should instruct the group to do several more trials.

Five What is the mean, mode, and range of the data? What do the data tell you?

Six Why do you think you have a reaction time, versus reactions being instantaneous? Is your body using its available energy during that time?

Seven How much time does it take for one person to react? How accurate is that time? How do you think reaction time affects an energy analysis of the human organism?

THE BICYCLE

Materials

- Bicycle
- Meter sticks
- Two spring scales
- String

Procedure

One Turn the bicycle over and balance it on the seat and handlebars.

Two Place the pedals so that they are parallel to the ground. Attach one spring scale to a pedal, and record the force (F_{in}) as you pull that pedal slowly towards the ground—be sure you are making the back wheel turn—until the pedal has moved half the distance towards being perpendicular to the ground, or 45°.

Three Return the pedals to the parallel position. Tie a piece of string around the tire. Attach the second spring scale to the string, so that it is parallel to the ground and pointed towards the front of the bicycle. Repeat Step two, applying the same force, and record the force (F_{out}) from the second spring scale. Make certain you turn the pedals in the same direction, and the bike is in the same gear, for Steps two and three. (See Figure 3.1.)

Four Measure d_{in}, the distance the pedal moves, and d_{out}, the distance the bicycle travels per one revolution of the pedal. Because the pedal moves along the path of a circle, d_{in} can be determined by finding the circle's circumference. Measure the radius of the pedal from the center to the pedal tip, and calculate the circumference ($2\pi r$). Measure the distance the bicycle travels, d_{out}, by rolling the bicycle forward so one peddle moves through one complete circle, and measuring how far the front wheel travels with a meter stick.

Five Calculate the work/energy input and output.

Six Calculate the efficiency of your bicycle.

ENERGY PICTURE OF A PERSON RIDING A BICYCLE

Procedure

One Using your results from these three investigations, describe how energy transfers from a food item, through a bicycle rider, to a bicycle. In doing so, you will need to address how reaction time affects energy transfers—do you need to supply energy for "thinking" or is it accounted for in the efficiency of the human body.

Two Based on your results, how much work a bicycle can do for 1 Calorie of food eaten by a cyclist? Calculate your answer in Joules.

Three Designate one member of your group to share these results with the class. Your teacher will provide a format.

Thinking and driving

Driving takes a lot of concentration and quick reaction times. Unfortunately, thousands of people die each year because they become impaired or distracted and their vehicle leaves the road.

Scientists from Carnegie Mellon University and AssistWare Technology Inc. have created RALPH (Rapidly Adapting Lateral Position Handler), a system that both warns drivers when their vehicle leaves the road and controls the vehicle's lateral position to keep it in its lane.

What kinds of technologies might go into this kind of system? What human functions would it aid?

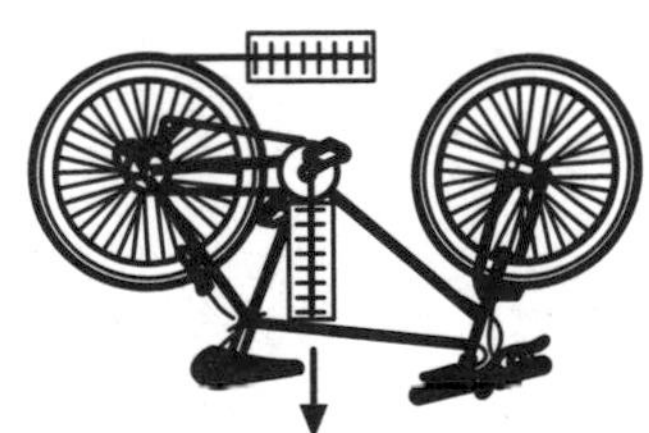

FIGURE 3.1
Bicycle and spring scales.

TEACHER-TO-TEACHER

The three investigations in this activity help establish the interdisciplinary nature of science, and may take one to two periods each depending how you use them. They require students to think of their bodies as machines—especially concerning the need for fuel. If students do not see the significance of the "energy to think and react," ask them if they need to keep energy flowing into a computer in order for them to work with a spreadsheet or search the Internet. You might also ask if any students have noticed a decrease in their ability to think and react if they have not eaten for a while. Remind them of how important reaction time is for safe driving.

Teacher's note

Of these exercises, the student group chosen to do "The Bicycle," should be made up of 4 responsible students for safety purposes. "The Bicycle" should probably be done first while the rest of the class observes, or works on "Food-to-Work." "Food-to-Work" may provide the best results if each student analyzes their own food wrapper, but works in a group with two or three others. "Human Reaction" will work best with one or two large groups of students.

Conceptual development

- Food provides the human organism with energy that is transformed into life functions, including thinking and mechanical work.
- Food energy is measured as available heat in Calories, whereas mechanical energy is calculated as force acting over a distance and measured in Joules.
- Energy transfers are never perfect, or 100 percent efficient. Therefore, the efficiency (ratio of output to input) of the many energy transfers involved in riding a bicycle can determine just how much work you could do from one food item.

Materials

- Food wrapper with nutritional information: Each student should bring their own to the first day of this activity.
- A bicycle: A stationary bike will also work, and does not need to be turned over.
- Meter sticks
- Spring scales: Two 20 N spring scales. If you have spring scales calibrated in grams or kilograms, you can make covers for those calibrations with the equivalent in Newtons (1 kg weighs 9.8 N).
- Human biology resources: Biology, or anatomy and physiolgy textbooks.

Teacher's guide

FOOD-TO-WORK

One Any food item will work, provided it has a nutritional label. Students should staple their wrappers into their journals. Answers will vary depending upon food items. For example, for an item having 150 Calories:

$$E_{in} \text{ (in Calories)} = 150 \text{ Cal}$$

Two Students should recognize that the digestion process begins the food-to-fuel process. The degree to which they elaborate on this will vary based on the amount of time, and what resources you provide for this exercise. You may choose to limit the in-class research to a basic description of digestion, and assign a more elaborate take-home assignment. (See Extension three.)

Three Students should base their answers on their experience, and their interpretation of human efficiency. A student might conclude that humans are efficient because he/she can play a full game of soccer after having eaten a light breakfast. Or, a student may decide that humans are not efficient because some companies have replaced human workers with machines. Students should provide reasons for their answers, and should demonstrate an understanding that greater efficiency means more output per input. (See Extension four.)

Four Efficiency is a ratio of output to input. Encourage your students to determine this formula for themselves, but provide it if necessary. Answers will vary depending upon food item. For example,

Efficiency= Output / Input

$\text{Efficiency}_{human} = E_{out} / E_{in}$

$E_{out} = (\text{Efficiency}_{human})\ (E_{in})$

$E_{out} = (.10)\ (150\ \text{Cal})$

$E_{out} = 15\ \text{Cal}$

Five Answers will vary depending upon food item. For example:

$E_{out} = (15\ \text{Cal})\ (4{,}180\ \text{J} / 1\ \text{Calorie})$

$E_{out} = 6{,}270\ \text{J}$

HUMAN REACTION

One Designate a student to control the stopwatch and another to record data. You may need to monitor this activity, and help with organization.

Two If students are uneasy about holding hands with a member of the opposite sex, create a circle with boys on one side and girls on the other, so that there are only two points at which a boy and a girl are holding hands. Do not create an all-boys and all-girls circle, as this may lead to unfair comparisons of, and unwanted criticism about the difference between the reaction times of the sexes.

Three If students question the need for repetition, you might want to explain how more data provide a clearer, and more accurate, picture.

Four See sample data table below (Figure 3.2).

TIMES TO REACT/TRANSFER FROM HAND TO HAND

	TRIAL 1	TRIAL 2	TRIAL 3	TRIAL 4	TRIAL 5	AVERAGE	# IN GROUP	INDIVIDUAL TIME
REACTION TIME	8.91	7.81	6.50	6.84	6.13	7.24	29	0.25

FIGURE 3.2
Sample data table for human reaction.

Five Numeric answers will vary according to these definitions: mean is the average, 7.24; mode is the most repeated value, not applicable in this data set; and, range is the difference between the largest and smallest values, 2.78. The significance of this data will vary for each student, but they should provide reasons for their statements and conclusions.

Six The cause of reaction time is the body's need to think: generate nerve impulse, process signals, decide upon reaction(s), send signal to muscle(s), and physically react. The human body must use energy during reaction time to stay alive and to think. Depending on class time, you might want to highlight the different types of reactions in the human body. There are involuntary responses, caused by the autonomic nervous system, such as digestion, and voluntary responses caused by the somatic nervous system, such as squeezing somebody's hand.

Seven Divide the average time by the number of students in the circle. This time is accurate within the parameters of this investigation. Students should conclude that some energy needs to be transferred into "thinking," in order to perform some mechanical tasks. An energy analysis is affected in that the students cannot measure this energy, and therefore cannot include it in their calculations.

THE BICYCLE

One Establish a safe area for this exercise. Determine that students understand the procedure and emphasize the need to measure in the correct direction.

Two Sample data:

$$F_{in} = 6.00 \text{ N}$$

Three Sample data:

$$F_{out} = 2.00 \text{ N}$$

Four Sample data:

$$d_{in} = 1.13 \text{ m}$$
$$d_{out} = 3.05 \text{ m}$$

Five Encourage your students to determine this formula themselves, but provide it if necessary. Sample data:

$$W_{in} = E_{in}$$
$$= (F_{in})\ (d_{in})$$
$$=(6.00 \text{ N})\ (1.13 \text{ m})$$
$$= 6.78 \text{ Nm (Newton-meters)}$$
$$W_{out} = E_{out}$$
$$= (F_{out})\ (d_{out})$$
$$=(2.00 \text{ N})\ (3.05 \text{ m})$$
$$= 6.10 \text{ Nm (Newton-meters)}$$

Six Sample data:

$$\text{Efficiency} = \text{Output} / \text{Input}$$
$$\text{Efficiency}_{bicycle} = W_{out} / W_{in}$$
$$\text{Efficiency}_{bicycle} = 6.10 \text{ Nm} / 6.78 \text{ Nm}$$
$$\text{Efficiency}_{bicycle} = .900$$

ENERGY PICTURE OF A PERSON RIDING A BICYCLE

Based on time, or other potential factors, decide if you want the class to work as a whole or in individual groups, and how you want the final analyses displayed. If there is not enough time, this section could be assigned as homework.

One Answers will vary.

Two Sample data:

$$Efficiency_{human} = E_{out} / E_{in}$$
$$E_{out} = (E_{in})\ (Efficiency_{human})$$
$$E_{out} = (1\ Cal)\ (.10)$$
$$E_{out} = .10\ Cal = 418\ Joules$$

$$Efficiency_{bicycle} = W_{out} / W_{in}$$
$$W_{out} = (W_{in})\ (Efficiency_{bicycle})$$
$$W_{out} = (418)\ (.90)$$
$$W_{out} = 376\ Joules$$

Three Have each group choose a spokesperson to present their findings to the class. Choose a method—chalkboard, overhead, computer display—for them to use.

Extensions

Extension one If your students cannot use bicycles, select any common, human powered vehicle that is real to them and re-write that section of this learning plan accordingly. There should be some way to measure the human force input and the vehicle force output. This extension can also be used as an extra assignment.

Extension two As noted earlier, the "Food-to Work" section of this activity could provide a lengthier assignment. For example, each student group could be assigned a digestive organ as the subject of an in-depth report focusing on its role in the food-to-energy process. This could then be made into a presentation. Assignments can also be designed for individual students, such as a report on energy and the human body—topics could include how certain diets effect energy levels, how a body suffering from a fever uses energy differently from a healthy body, how different levels of activity result in a body using more or less energy. Assignments will vary greatly depending on your curriculum and the abilities of your students, but each should be designed to provide a definitive link between energy and human biology.

Extension three Engage your students in a discussion, or design a research topic that relates the human body to that of an automobile. For example, both need energy to function, and both use more energy when operating in strenuous or extreme conditions. Both need to be refueled, periodically, and both perform better on higher quality fuels. Research will show that the simple human system, a "glucose eating machine," is more efficient than the simple internal combustion engine.

Extension four What is efficiency? Ask your students if "best" and "most efficient" are interchangeable terms.

Extension five Can efficiency be improved? As a take-home assignment, ask your students to research a particular design that has been improved upon with the goal of obtaining better efficiency. For example, cars—aerodynamic shapc, lighter materials, better fuels; bicycles—lighter materials, stronger frame design. Students can also research the human body to learn how exercise and training can improve human efficiency.

Activity 2

Energy of motion

While riding a bicycle, do you have energy simply due to your motion?

Challenge

What determines how much energy of motion you can generate? How do you, and your attributes, factor into this equation? What factors determine stopping distance? How much energy does stopping require? How much energy of motion—kinetic energy—do you possess?

Objectives

- To discover and recognize the factors of kinetic energy.
- To calculate the kinetic energy of a bicycle at various speeds.
- To determine the relationship between speed and kinetic energy.
- To generate a hypothesis about the work needed to stop, as well as its factors.

Materials

- Calculator or computer
- Scale calibrated in Newtons

Procedure

One Measure your weight on a scale that has been calibrated in Newtons.

Two Calculate your mass; 1k weighs 9.8 Newtons at sea level on Earth.

Three Use your mass and the following formula to create a table showing your kinetic energy if you were riding a 15 kg bicycle at the following speeds (all are in meters per second): 2.2, 4.4, 6.7, 8.9, 11.1, 13.3, 15.6, and 17.8.

$$KE=(\tfrac{1}{2})(m_{total})(v^2).$$

Four Create a graph to determine the relationship between kinetic energy and speed.

Five What is the relationship between speed and kinetic energy?

Six What is the relationship between mass and kinetic energy?

Seven How much energy of motion do you have when riding a bicycle? What is the significance of this amount in relation to transportation safety?

Eight In order to stop, you must do at least the same amount of work as you did to speed up. What, then, is the relationship between your speed and the work needed to stop at that speed?

Nine Assume that the work needed to stop you is the product of the frictional forces and stopping distance:

$$W_{stop} = (F_{frictional})(d_{stopping})$$

Based on this assumption, what is the relationship between your speed and your stopping distance (assuming $F_{frictional}$ is about the same for all speeds)?

Ten While riding a bicycle, do you have energy simply due to your motion? Why, or why not?

■ TEACHER-TO-TEACHER

"Energy of Motion" allows students to draw a real life connection between human biology, physics, and transportation safety. The goal is to portray a tangible relationship between mass, speed, kinetic energy, and stopping distance. In doing this, students will have the opportunity to build the language of mathematics within the context of science, and to use spreadsheets and graphs to discover trends in data.

This activity can also be used to either introduce your students to spreadsheets, or hone existing skills. It will take one to two periods.

Conceptual development

- Kinetic energy is directly related to a vehicle's mass and, more importantly, exponentially related to speed—two times the speed equals four times the kinetic energy.
- The work done to start and stop a vehicle is directly related to its kinetic energy, therefore directly related to its mass.
- Several energy transfers occur when you ride a bicycle, including that of work into motion.

Materials

- Calculators or computers: Determine whether your students will do this activity with a calculator, or if you want to use spreadsheets.
- Scale calibrated in Newtons: If you are unable to make this calibration, most bathroom scales are calibrated in both kilograms and pounds. Students can do their conversions backwards to determine what their weight is in Newtons.

Teacher's guide

One Weights will vary, but a "nice" average is 500 N. The following sample data will all be based on that value. It is imperative that students understand the difference between weight and mass for this activity.

$$\begin{aligned} \text{Weight} &= F_g \\ &= 500\text{ N} \end{aligned}$$

Two

$$\begin{aligned} m_{yourself} &= F_g\,(1\text{ kg} / 9.8\text{ N}) \\ &= (500\text{ N})\,(1\text{ kg} / 9.8\text{ N}) \\ &= 51\text{ kg} \end{aligned}$$

Three Tables will vary, but should include all of the following information, and values should be appropriately labeled. (See Figure 3.3.) Allow students to determine their own scale. The following pattern will emerge, regardless of scale, unless the student made mathematical errors or mislabeled the axes.

Four Allow students to determine their own scale. The pattern in Figure 3.4 will emerge, regardless of scale, unless the student made mathematical errors or mislabeled the axes.

Five There is an exponential relationship between speed and kinetic energy. Students' graphs should show a parabolic curve and, depending on their mathematical background, students may recognize that this curve represents an exponential relationship.

Six Students should determine that mass is directly related to kinetic energy, a relationship expressed by the mathematical formula for kinetic energy. Students should note that this means that more massive objects will have more kinetic energy.

Seven Answers will vary, but all students should state that their energy of motion will depend on their mass and speed. Students might conclude that the kinetic energy of an object plays a large part in determining how much damage it will do when it impacts with another object. (This concept is developed further in "Collisions and Safety," beginning on page 121.) If some students do not see this, point out that both kinetic energy and force are measured in Newtons. Once all students understand this relationship, you might choose to discuss transportation safety, especially that related to driving slow. (See Extension one.)

CONCEPT	METRIC SPEED	YOUR MASS	BIKE'S MASS	TOTAL MASS	KINETIC ENERGY
SYMBOL	v	m_{yours}	$m_{bike's}$	$m_{total} = m_{yours} + m_{bike's}$	$KE = (½)(m_{total})(v^2)$
UNITS	m/s	kg	kg	kg	Newtons (N)
VALUES	2.2	51	15	66	159.72
	4.4	51	15	66	638.88
	6.7	51	15	66	1481.37
	8.9	51	15	66	2613.93
	11.1	51	15	66	4065.93
	13.3	51	15	66	5837.37
	15.6	51	15	55	8030.88
	17.8	51	15	66	10455.72

FIGURE 3.3
Sample data table for Step four.

Eight Since the work needed to stop is directly proportional to kinetic energy, and speed is exponentially proportional to kinetic energy, work needed to stop is exponentially proportional to speed.

Nine As with the relationship between speed and kinetic energy, students should be able to find an exponential relationship between speed and stopping distance.

Ten Yes. Kinetic energy is solely the result of speed.

Extensions

Extension one After discussing the relationship between speed, mass, and kinetic energy, take the opportunity to relate these results to other types of motion your students have experienced, such as automobiles. Ask each student to find the mass of their parents' vehicle(s); this should be available in the operator's guide, on a sticker in the driver's door, or students can call a dealership. You might also ask for the mass of their school bus. As either an in-class or homework assignment, have them determine the kinetic energy of these vehicles during average situations: 15 kph in a school zone, 25 kph in a residential area, 35 or 45 kph in a town or city, and 55 or 65 kph on the highway.

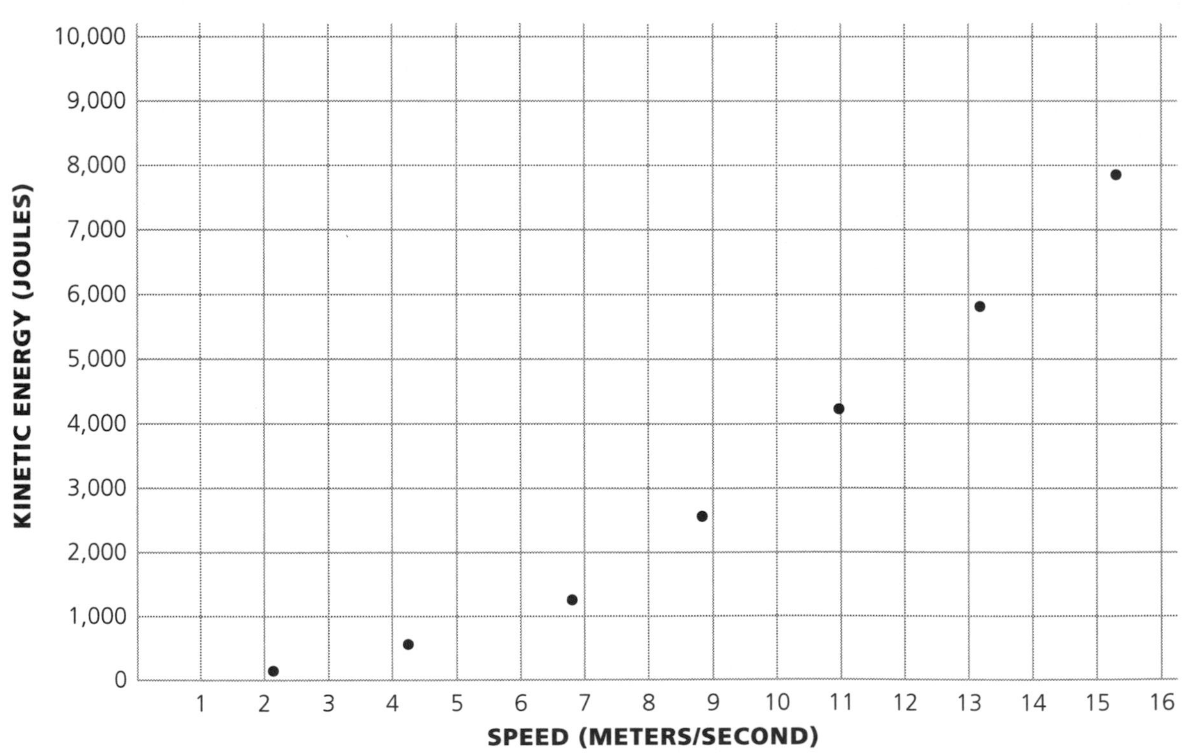

FIGURE 3.4
Sample data table for Step five.

Discuss their results in terms of safety. You might discuss energy transfers, what happens when a vehicle collides with another object, which will be covered in Activity 3 "Energy Transfers." Review what they have learned about stopping times, and discuss the rationale of slow speeds in highly populated areas.

Extension two Students may want to measure their actual speeds on a bicycle. If your school has an exercise bike with a speedometer, or if you have access to a bicycle with a speedometer, students can use them to calculate kinetic energy. They can then compare this field data to that gathered in class. A comparison of the two separate experiments could be used as a take-home assignment.

Extension three The fundamentals of this activity can be applied to other motion-related sports. Students might enjoy determining the kinetic energy of athletes competing in a wide variety of sporting events and situations. Students might also enjoy finding the kinetic energy of inanimate objects, such as pucks and balls, and sticks, bats, and clubs. Statistics—mass and average speed—for athletes and their equipment should be available through various websites, newspaper articles, and books.

Activity 3

Energy transfers

How does energy transfer through a vehicle?

Challenge

How does the energy in the food you eat enable you to lift a ball? Is energy transferred from you to the ball? Can you follow that energy as you throw the ball? When it eventually comes to rest? How much energy is involved in those transfers? Is the energy transferred efficiently? Where does the energy go when the ball stops?

Objectives

- To calculate the potential energy in work done when lifting an object.
- To explore the relationship between potential and kinetic energy.
- To explore the relationship between kinetic energy and stopping distance.
- To describe energy transformations in a system containing both biological and mechanical components.
- To calculate the efficiency of a system containing both biological and mechanical components.

Materials

- 30 cm ramp with center groove
- 100 cm ramp with center groove
- metal ball
- pieces of wooden dowel to stop ball.
- meter stick
- scale or balance

Procedure

One Find the mass of the ball you will use.

Two With your group, use all the materials provided to set up a system similar to the illustration at the beginning of the activity. Place the wooden dowel in the groove of the 100 cm ramp near the bottom of the 30 cm ramp at the beginning of each trial. Using books as a prop, find a height from which the ball can be released and not roll off the end of the 100 cm ramp. Use several test runs to determine this height.

Three Run several trials at the height you established in Step one (or lower if necessary) recording the distance the ball travels— $d_{stopping}$— each time. When you can repeat the same, or very similar, stopping distance several times, record that stopping distance and height.

Four How much work did you do to lift the ball? Use your data and the following formula to determine the potential energy of the ball—its energy at place #2.

$$\begin{aligned} Energy_2 &= w \\ &= \text{Potential Energy} \\ &= (F_{up})(h) \\ &= (m)(g)(h) \end{aligned}$$

where $g = 9.8\ m/s^2$.

What must you do to give the ball more energy?

Five How much food would you need to eat in order to lift the ball? Use the potential energy of the ball, a human efficiency value of 10 percent, and the following formula to calculate this food energy, $Energy_1$, in both Joules and Calories.

$$\Sigma\ Efficiency = \frac{output}{input} = \frac{Energy_2}{Energy_1}$$

Six What happens to the potential energy of the ball as it rolls down the ramp? The potential energy of the ball is transferred into kinetic energy as it rolls down the ramp, so that at the bottom the ball has only kinetic energy. This transfer is never 100 percent, but imagine it is for the purposes of this activity.

The wooden dowel provides the largest frictional force that stops the ball, much like the interaction between the road and the breaks and tires of an automobile. How much force must it apply to stop the ball? Using the stopping distance measured in Step one, calculate the force the wooden dowel must have applied against the ball ($Force_{friction}$).

$$\begin{aligned} Work_{stopping} &= (F_f)(d_{stopping}) \\ &= Energy_3 \end{aligned}$$

Seven As the ball stops, all of its kinetic energy is transferred into heat. For the purpose of this activity, assume all this heat went into the ball and calculate this change in temperature. (Your teacher will provide c, the specific heat of the metal.)

$$\begin{aligned} Energy_4 &= Energy_3 \\ &= Heat = (m)(c)(\Delta T) \end{aligned}$$

Discuss the results of your calculations with your group. What do they mean? Record your impressions.

Eight What is the relationship between

Fuel efficiency

The race has been on for a while to design a more fuel efficient and environmentally friendly vehicle. In response to these concerns, scientists in the field of automotive technology have created the Hybrid Electric Vehicle (HEV). This vehicle uses the same engine and fuel tank as the vehicles currently being used, but combines them with an electric motor. The practical benefits of HEVs include improved fuel economy and lower emissions compared to conventional vehicles and the capability of using petroleum or alternative fuels.

What part of your body would you modify to increase your energy efficiency? What changes could you make to your bike or skateboard to increase their efficiency?

height and stopping distance? Perform 12 more trials, three at each of four different heights, and measure the stopping distances. Record your data in a table, and plot a graph of height versus average stopping distance.

Nine Describe the relationship you have discovered between height and stopping distance, and explain it in terms of the theoretical energy analysis you performed in Steps three, five, and six of this activity.

■ TEACHER-TO-TEACHER

Besides being a theoretical walk through of an energy analysis, this four-period activity equips students with tools to calculate the energies involved in colliding objects, from football players to automobiles. This activity is intended as an analogy between an experiment students can perform, and an actual vehicle situation that cannot be studied in the classroom. In performing this activity, students can develop their logic and general thinking skills as they move through each step. Students will explore the relationship between height and stopping distance, and extrapolate their results to the relationship between the speed and stopping distance of a vehicle traveling on a road. Students should also develop a better understanding of the notion of energy forms and transfers, as well as the very important transportation issue of stopping distance.

To illustrate the idea of energy transfer, place a Newtonian Demonstrator—called Newton's Cradle in commercial stores—on your desk before beginning the activity. Start by lifting one ball, and letting it go for a while, then lift two. As students watch, ask them to explain what they think is happening.

Conceptual development

- Energy can be transformed from food to work, or potential to kinetic.
- These transformations can be analyzed, producing data which reveal relationships that cannot be observed in a simple force analysis.
- Energy analyses of small systems can be used to study analogous systems that might be too large to reproduce in a laboratory, and can illuminate the ideas involved in complicated transportation situations.

Materials

- 30 cm ramp with center groove and mechanism to support it: A plastic ruler will work well for this ramp and the support mechanism is most easily made from students' books.
- 100 cm ramp with center groove: This can be a plastic toy track, a piece of "U" or "H" shaped metal, or several plastic rulers.
- Metal ball: A large marble-sized ball that is made of a metal with a known specific heat is best, but is not absolutely necessary. If you use a ball with an unknown specific heat, assign a specific heat which will result in a believable change in temperature. It is important only that students understand that some energy in an energy transfer is lost to heat, and that the amount can be measured.
- Pieces of wooden dowels: You may need to cut up pieces of the same materials that will fit the groove of the 100 cm track you use. You should sand the ends of the dowels to avoid splinters and help create a constant frictional force while stopping the ball.

- Meter stick: Each group will need at least one.
- Balance or scale: Students will need to mass the balls they use for this activity.

Teacher's guide

One You might want to inform students that they will only receive one ball for this investigation. If they lose the ball, you might "make up" data for them. Students should have no difficulty determining the mass of the ball.

Two Set-ups will vary depending upon available equipment, but should be similar to the diagram provided. Be aware of the time as students determine the height they want to use.

Sample Data:

$m = 66.2\text{ g} \quad = 0.0662\text{ kg}$

Three Sample Data:

$h = 8.3\text{ cm} \quad = 0.083\text{ m}$

$d_{stopping} = 13.9\text{ cm} = 0.139\text{ m}$

Four Students must lift the ball higher, or use a ball of greater mass to give the ball more potential energy.

Sample Data:

$Energy_2 = W$
$= \text{Potential Energy}$
$= (F_{up})(h)$
$= (m)(g)(h)$

$Energy_2 = (0.0662\text{ kg})(9.8\text{ m/s}^2)(0.083\text{ m})$
$= 0.054\text{ J}$

Five Sample Data:

$Energy_1 = (Energy_2)(Efficiency_{human})$
$= (0.054\text{ J})(0.10)$
$= 0.00545\text{ Joules}$

$Energy_1 = (\text{ J})(1\text{ Calorie} / 4{,}180\text{ J})$
$= (0.00545\text{ J}) / (4{,}180\text{ Cal/J})$
$= 1.3 \times 10^{-6}\text{ Cal}$

Six The ball's potential energy becomes kinetic energy as it rolls down the ramp. The work from friction elliminates the kinetic energy.

Sample Data:

$Work_{stopping} = (F_f)(d_{stopping})$
$= Energy_3$

$F_f = (Energy_3) / (d_{stopping})$
$= (0.054\text{ J}) / (0.139\text{ m})$
$= 0.34\text{ N}$

Seven Although the energies involved are small, the change in temperature indicates that some energy was lost as heat.

Sample Data for iron (c = 0.109):

$Energy_4 = Energy_3 \times (1\text{ Cal} / 4{,}180\text{ J})_4$
$Energy_4 = 1.3 \times 10^{-6}\text{ Cal}$
$Energy_4 = (m)(c)(\Delta T)$

$\Delta T = Energy_4 / (m)(c)$
$= (1.3 \times 10^{-6}\text{ Cal}) / (0.0662\text{ kg})(0.109)$
$= 1.8 \times 10^{-3}\text{ °C}$

Eight Answers will vary.

Nine The stopping distance is directly related to height and, as height increases, so does stopping distance. Height is directly related to potential energy; potential energy is directly related to kinetic energy; kintetic energy is directly related to the work necessary to stop; and the work needed to stop is directly related to the stopping distance. These relationships are shown in the following equation:

$$\begin{aligned} \text{Potential Energy} &= (F_{up})(h) \\ &= (m)\ (g)\ (h) \\ &= \text{Kinetic Energy} \\ &= \text{Work}_{stopping} \\ &= (F_f)\ (d_{stopping}) \end{aligned}$$

Therefore h is directly proportional to $d_{stopping}$.

Extensions

Extension one Set up a more realistic "vehicle" by performing a similar analysis using a track large enough for the toy truck from "Balanced Vehicles" (page 44). Or, use a car and track set, if one of your students has one to lend.

Extension two This activity assumes that all potential energy becomes kinetic energy, and that all of the energy lost to heat is absorbed by the ball. Remind students this is not the case in real energy transfers, and ask them to create an actual energy analysis of an energy transfer situation. You may choose to have students research their analysis, or hypothesize based on what they have already learned.

Extension three Increase the precision of this activity by directly measuring all the parameters involved. Measure the forces with spring scales, use photogate timers to calculate the speeds, and probeware to measure temperature increases.

Extension four If your students enjoy these energy analyses, you could challenge them to make Rube Goldberg Machines. These are whimsical devices that do simple tasks in quite complicated manners. They are fun to build, analyze, and present to the class.

Activity 4

Speeding up and stopping

How does the energy transfer from its source to the motion of a vehicle? What makes a vehicle go?

Challenge

When you are driving a vehicle, you are transferring energy from one source to many places. What is the energy source, and what are the transfers? How much energy does it take to reach and maintain a certain speed? How much energy does it take to stop? Does speed affect that amount? Where does energy go when you stop?

Objectives

- To research the energy transfers in a vehicle.
- To perform an energy analysis of a vehicle.
- To apply knowledge gained in previous activities.
- To present a vehicle energy analysis to your class.

Materials

- Vehicle information resources

Procedure

One In your groups, decide which vehicle you will analyze. Consider your familiarity and/or interest while choosing your "favorite" vehicle. Possible vehicles include bicycles, skate boards, cross country skis, motorcycles, various cars and trucks, buses, airplanes, and trains. When choosing a vehicle, consider your available resources. Draw and describe your vehicle in your journal.

Two Describe the energy source and transfers for your vehicle in these three situations: speeding up, maintaining speed, and stopping.

Three Document all your sources of information.

Four Prepare a written and oral presentation of your energy analysis and present it to your peers.

■ TEACHER-TO-TEACHER

This activity is the culmination of what students have learned in Activities 1, 2, and 3. It was designed to solidify the connection between classroom physics, transportation, and transportation safety. It will take three to four periods.

Conceptual development

- Theoretical science can be applied to any situation.
- By understanding the basic ideas behind fuel, energy transfers, and motion, students gain an understanding of everyday transportation situations.

Materials

- Vehicle information resources: Internet access, magazines. Students may visit a garage or bicycle store for first hand information.

Teacher's guide

One You may wish to develop a particular format for describing their chosen vehicle in order to standardize information between the groups.

Two Students should discover that the energy used to speed up is greater than the energy needed to maintain a speed. The energy involved in stopping is a theme that runs throughout the "Energy and Detection" chapter, and is included here for that reason.

Three The quality of their information is very important, so you should demand accountability before presentations are given. Encourage students to interview garage or bicycle store mechanics and, if they do, to include their source's name, certification/experience, and place of employment. (See Extension one.)

Four Provide students with presentation criteria to insure the students know your expectations before they prepare their presentations. These presentations could be designed as a science showcase for principals, parents, and other classes.

Extensions

Extension one In addition to having students collect data outside of the classroom, you might choose to organize a field trip to a local driving school, or automobile testing site. If this is possible, the experience will provide first hand data for students, as well as a different perspective on driving.

Extension two Have students make recommendations either for or against a particular vehicle. Students should base their decisions on the presentations and cite them accordingly.

Background reading

Energy analysis

An energy analysis is a step-by-step description of each energy transfer involved in a situation. For example, an energy analysis can be used to determine how fast you could ride your bike based on the amount of food you eat before you ride. This type of analysis reveals more information than force or motion analyses, and can be used to study a separate situation altogether.

For example, consider analyzing the energy required to lift a ball up a ramp, then the energy released when it rolls down and comes to a stop. We can make an analogy between the motion and energy of this ball and a person riding a bicycle. The energy contained in food is transferred into and through the human organism by the processes of eating and digestion. The energy is then imparted to the bicycle as mechanical energy through pedaling, analagous to lifting a ball up. A rider can just roll down a hill, letting the force of gravity keep him/her going, analagous to the ball rolling down the ramp. Finally, a rider uses the brakes to stop a bicycle on a horizontal plane, similar to the ball stopping on flat ramp. Let's examine an energy analysis of a ball/ramp situation.

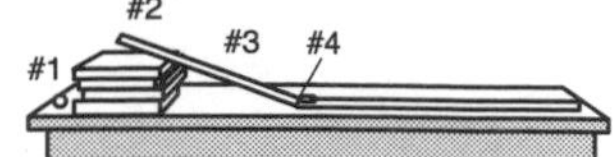

FIGURE 3.5
Ball-ramp analogy.

A ball (solid, round, circular vehicle) is lifted up (F_{up}) to a spot on a ramp (h) by a human doing work. It then rolls down the ramp to a speed. On the horizontal ramp friction (F_f) acts over a stopping distance ($d_{stopping}$) to bring the ball back to rest.

"$\Delta Energy_{1 \text{ to } 2}$" is the change of energy from place #1 to place #2. The source of the energy is you. You can calculate how much food you need to do that much work by converting the work (Joules) into Calories of heat and factoring in human efficiency.

"$\Delta Energy_{2 \text{ to } 3}$," is the change of energy from place #2 to place #3. The potential energy of the ball is converted to kinetic energy as it rolls down the ramp (or falls at an angle). Note that the equation implies that all the potential energy is converted into kinetic energy, even though that never happens. In reality, some energy is lost to heat through friction.

"$\Delta Energy_{3 \text{ to } 4}$" is the change of energy from place #3 to place #4. The kinetic energy of the ball is lost to heat as the frictional force (F_f) does work to stop the ball over the stopping distance ($d_{stopping}$).

"$Energy_1$ = source" means that the energy that transfers to the ball while lifting comes from some source. At place #1, the ball has zero energy—no place to fall, thus no gravitational potential energy, and no motion, thus no kinetic energy.

"$Energy_2 = W = (F_{up})(h) =$ Potential Energy" means that at place #2, the ball's energy is the work (W) you did on it to lift it up with a force (F_{up}) to a height (h). Because the ball can not fall, we would say that all of the work went into potential energy.

"$Energy_3 =$ Kinetic Energy $= (½)(m)(v^2)$" means that the energy at place #3 is now energy of motion, or kinetic energy. Kinetic energy is directly related to the mass and speed of the ball.

"$Energy_4 =$ Heat $= (m)(c)(\Delta T)$" is the energy at place #4. All the mechanical energy (work, potential, and kinetic) are transformed into heat, presumably through friction. We can find the amount of energy lost to heat (ΔT) if we know the specific heat (c) of the material of the ball.

Work, Kinetic Energy, and Stopping

Work is the primary measurable concept of mechanical energy. The basic unit for energy is derived from the idea of work. When 1 Newton (N) of force (F) acts over 1 meter (m) of distance (d), 1 unit of work (W) has been performed. That 1 Newton-meter (N-m) of work is defined as 1 Joule (J). Work is defined in physics as occuring only when a force acts over a distance, and both the force and distance are in the same direction. It is important to note that, in everyday language, work can mean many things, including homework. If you are reading this as homework, are you doing any work as defined by physics? Perhaps if you were taking notes, the force pushing the pen would be doing work. Even though you may be transferring lots of energy, like you do when thinking, no "real" work is being done.

When a force acts to speed up an object, we say that work is transferred to kinetic energy. In math, we might write $W = \Delta KE$ to symbolize that work has caused a change in kinetic energy. This situation, in which a force acting over a distance to speed up an object, also gives us the precise definition of kinetic energy, $KE = ½(m)(v^2)$. But, where does the ½ come from? Why v^2 ? If you were to work out the theoretical force and motion analysis, you would find that as a force speeds up an object, the object covers more distance in each consecutive second-more distance is covered in the second second than was covered in the first second. Therefore, the distance relationship to speed is exponential, and not linear.

Since distance is directly related to work, or energy, kinetic energy needs to be related to the square of speed to be consistent with the exponential relationship between distance and speed. Also, to be consistent with the definitions of force, speed, and acceleration, average speeds during each second must be used to determine the distances traveled while accelerating. Those average speeds give us the factor ½. (For the purposes of this book, it is more important that you understand that there is a relationship between speed, mass, and kinetic energy, than that you know how to explain it. (Acceleration, the rate at which an object speeds up, is not related to work or kinetic energy, and is a subject covered in higher levels of physics.)

When considering the work required to stop a vehicle, all the relationships are the same as when speeding up. However, when driving an automobile, these relationship have serious consequences. Once on the highway, and traveling at greater speeds, we loose some of our ability to relate distances to speeding up and slowing down. Also, we have no natural way of knowing just how much kinetic energy our vehicle possesses at those speeds.

Large kinetic energies transferred over short distance involve huge forces that can do tremendous damage. To avoid such huge forces, we need to allow large distances for smaller forces to act. Thus, the exponential relationship between speed and distance to stop is very important. If it takes approximately 30 m to stop when traveling 50 kph, then at twice the speed, 100 kph, it will take four times the distance, or 120 m to stop. And at 100 kph, you travel those 120 m in just 1.08 seconds. It is difficult for us to plan ahead to allow enough distance to stop when traveling freeway speeds.

Heat, Temperature, and Units

What is **heat**? We can measure heat in degrees—Celsius, Fahrenheit, Kelvin—or Calories, but temperature is not matter. Changes in temperature can cause heat to move, which is somewhat analogous to our mechanical systems in which force causes objects to move. However, there is no object moving in this heat, or **thermal**, system. Heat is **thermal energy**, and **thermodynamics** is the theory that explains the motion of heat. It is also one of the essential theories in the physical sciences.

One way of developing a practical, conceptual understanding of thermodynamics, as well as other theories, is to consider how the units are defined. Scientists prefer to have a one-to-one relationship between cause and effect concepts and their units. In thermodynamics, a change of one degree in temperature will cause one unit of heat to move. Within the metric system, similar relationships exist between thermal, spatial, fluid, and some mechanical units, all of which are based upon water. This means that a 1°C temperature difference in water is related to 1 Cal of heat. The volume of water must be taken into consideration—more water needs more heat or thermal energy to heat it up—so we use a particular spatial measure for volume, 1 cubic centimeter. One cm^3 of water needs 1 Cal of heat to change 1°C in temperature. The cubic centimeter also defines our liquid volume and mechanical mass units. For example,

$$1 \text{ cm}^3 \text{ of water} = 1 \text{ ml of water}$$
$$1 \text{ cm}^3 \text{ of water} = 1 \text{ g of water}$$
$$1000 \text{ cm}^3 \text{ of water} = 1 \text{ L of water}$$
$$1 \text{ L of water} = 1 \text{ kg of water}$$

Therefore, 1 kg of water needs 1 kilocalorie, or 1 Cal, of heat to change temperature by 1°C. So, if you take 1 L or 1 kg of water at 0°C and add 1 Cal of heat, it would then have a temperature of 1°C. Keep adding heat, and the water temperature rises accordingly in a nice, simple relationship.

However, something very interesting happens when that liter of water gets to be 100°C. You can keep adding heat, but you will get no temperature change, until you have added 540 Cal. At that point, the water turns to a gas, and the temperature of that gas is still only 100° C. Water vapor is very much hotter than liquid water even when they are both 100°C. To get a H_2O molecule to go from a liquid to a gas you need a lot of thermal energy, exactly 540 Cal/kg or L. A similar situation occurs when you add heat to ice. Once the ice reaches 0°C, you will need to add 80 Cal/kg just to get 0°C water.

So, you can equate heat and temperature of water with a nice one-to-one relationship, 1 Cal to 1°C. However, you cannot say that the heat in 1 liter of water at 10°C is 10 Calories. To figure out how much thermal energy is in a kilogram of water at 10°C you would need to start with that mass of H_2O molecules at absolute zero, 0°Kelvin, at which there is no thermal energy. You would then add heat until you get water at 10°C.

Calories, calories, and Joules

Food Calories are really kilocalories. It makes some sense in the food sciences to simply drop the kilo and write a capital "C," but it is inconsistent with the rest of science. A kilometer is not called a Meter. Also, the units for energy in thermal systems and mechanical systems are conceptually unrelated. In a mechanical system when you apply a 1 N force over 1 m of distance, you have done 1 Joule of work. Thus the units for mechanical energy are related to the basic notion of work in a nice one-to-one manner A similar, howver, does not hold true for Joules and Calories, where 4.18 J = 1 Cal. or 4,180 J = 1 Calorie.

Specific Heat

Materials can be classified by how much energy, or heat, it takes to change their temperature. This physical property of a material is called **specific heat**, and is noted by the symbol c (not to be confused with the letter "c" which stands for the speed of light, as in the energy mass conversion formula $E = mc^2$). Materials with high specific heat values need a lot of heat to change temperature. These materials are good sources of heat, because they can "store" a lot of heat with little temperature change. Water has a high specific heat value. It is used in automobile radiators because it absorbs great amounts of excess heat and carries it away from the engine. In nature, oceans and lakes tend not to change their temperatures from day to night because those large bodies of water take much longer than 24 hours to transfer heat than does the land (rocks and soil) which has relatively low specific heat. Thus, people living next to large bodies of water experience smaller differences in the air temperature between day and night than people in inland areas.

Specific heat is measured in calories per gram, and water has a specific heat of 1. It takes 1 calorie to change the temperature of 1 g of water 1°C. Some common specific heats are as follows.

Water	1.000
Copper	0.092
Zinc	0.093
Iron	0.109
Mercury	0.033

Notice that mercury has one third the specific heat of water. This is one reason we use mercury in thermometers.

Symbols and Units

There are many seemingly random letters that are involved with science. (In order to distinguish symbols from abbreviated units in this particular section, **symbols** are in bold print.) These letters can be confusing, and it is very important for you to know someone is talking about the mass "**m**" of an object and not the distance "m" the object moved—3 g or 3 m. All ideas in science can be quantified by **symbols**. Scientists decide upon a symbol for an idea so that it can be referred to easily, and can be written the same way in any language. "Change," for example, is represented by the Greek letter Delta, Δ. Δ**T** means change in temperature, and Δ**v** means change in speed. But speed, an idea that is symbolized by **v**, can be represented by the units in which it is measured, kilometers per hour. Kilometers per hour, in turn, is often abbreviated to kph.

An abbreviation for a particular unit may be a single letter that is easily confused with a symbol. It is very important to be able to differentiate between symbols of an idea, and abbreviations of a unit. For example, the basic unit for distance is a meter, m. The concept of mass, however, is represented by **m** and measured in kilograms, kg. And, distance can be measured in many different ways, or ideas, and represented by different symbols, such as height (**h**), distance (**d**), width (**w**), length (**l**), or radius (**r**).

Science is notorious for using the same symbol over and over again, with subscripts denoting differences. If you wanted to describe the energies at several location, you could use $\mathbf{E_1}$ for the energy at location #1, $\mathbf{E_2}$ for the energy at location #2, and so on. The following chart may help you keep some of the ideas in classical mechanics organized. Copy this chart, and add to it as you continue to expand your knowledge of science.

CONCEPT	SYMBOL	DEFINITION	UNITS (ABBR.)	FORMULA DERIVATION
Work	**W**	force applied over a distance	Joule (J) (1 J = 1 N m)	W = Fd
Kinetic Energy	**KE**	energy of an object as a result of its motion	Newton (N) (1 N =1 kg m/s²)	KE = ½mv²
Force	**F**	causes change in the motion of an object	Newton (N) (1 N =1 kg m/s²)	F = ma
Force of Gravity	$\mathbf{F_g}$	force between two objects caused by their mass	Newton (N)	$F_g = mg$ (g =9.8 m/s² on Earth)
Acceleration	**a**	rate at which an object changes velocity	m/s²	$a = \Delta v/t$
Velocity or Speed	**v**	rate at which an object travels a certain distance *Note: this definition is for average speed*	m/s	v = d/t
Mass	**m**	measure of inertia	kilogram (kg)	
Distance	**d**	space between two events	meters (m)	
Time	**t**	interval between two events	seconds (s)	

FIGURE 3.5
Concepts and symbols.

chapter four

Detection

■ TEACHER-TO-TEACHER

This learning plan asks students to experience waves in a controlled environment, and build a visual base for modeling sound and light. Students are then asked to explore their ears and eyes, comparing them to examples of audio and visual technology and assessing their importance in transportation situations. Students then describe what they need to detect with these senses while operating several different vehicles. Finally, they create vehicle technologies to enhance a vehicle operator's ability to hear and see.

Conceptual development

Waves behave very differently than objects. Therefore, wave mechanics, the hows and whys of wave motion, is quite different than classical mechanics. The following paragraphs describe the concepts developed in the four activities of this chapter.

The human ear and eye can be analyzed as technological devices as well as living organs.

Humans gather information from their environment primarily though waves: mechanical waves like sound and electromagnetic waves like light. Humans have developed the ability to translate the energy of sound and light waves into information that the brain can decipher.

National Science Education Standards

In the development of the conceptual construction of this chapter, the following Content Standard concepts and principles served as a guide.

- Recognize and analyze alternative explanations and models (page 175).

Conceptual construction

Prior experiences of detecting (hearing, seeing, etc.) the environment while moving.

Central Idea Humans gather information primarily through waves.

Waves Explore wave phenomena in a coil spring.
Do You Hear What I Hear? Explore human ear and related technologies.
Picture This Explore the eye and related technologies.
Detecting While Moving In teams, present an analysis of vehicular detection.

Constructed and Communicated Idea Information from waves is detected to control vehicles.

- Formulate and revise scientific explanation and models using logic and evidence (page 175).
- Waves, including sound and… light, have energy and can transfer that energy when they interact with matter (page 180).
- Electromagnetic waves include… visible light… Each kind of atom or molecule can gain or lose energy only in particular discrete amounts and thus can absorb and emit light only at wavelengths corresponding to these amounts (page 180).
- In sense organs, specialized cells detect light, sound, and… enable animals to monitor what is going on in the world around them (page 187).

Time management

The entire learning plan may take eight fifty-minute periods. Activity 1 will take two periods, as students will need time to experience wave phenomena. Activities 2 and 3 may take two periods each, depending on the depth to which you choose to research each topic. Activity 4 may take two periods.

Learning plan assessment

In this learning plan, the fourth activity may not be indicative of what students have learned in the activities. You may want to assess student responses in terms of depth of thought, using questions such as:

- What did you learn about waves, and how you hear and see in these activities?
- Has what you learned in "Detection" changed your understanding of waves, and how you hear and see? How?
- How have these activities affected your knowledge of the operation of a vehicle?

In addition, students can perform peer reviews on each presentation. Create an assessment rubric (see Appendix, figure A-1) before the first presentation.

Activity 1

Waves

How are waves different from moving vehicles?

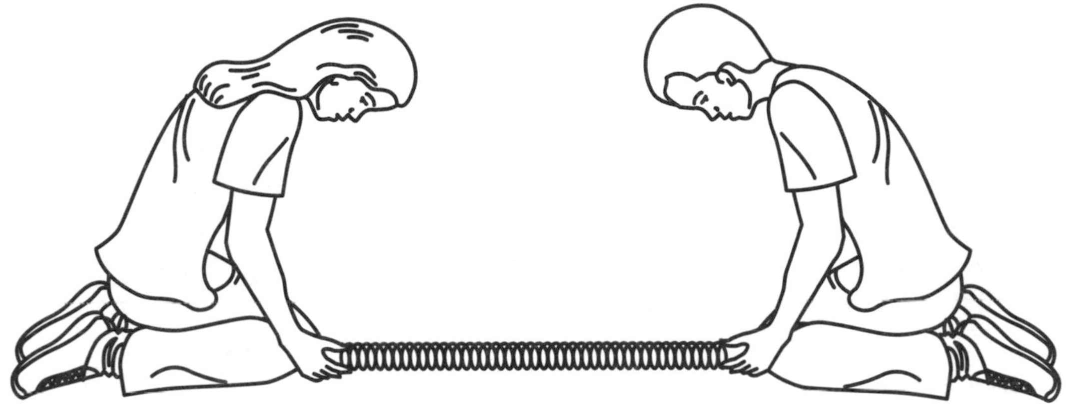

Challenge

How are waves important to our vision and hearing? Can waves be made to move faster or slower? Can two waves exist in the same place at the same time?

Objectives

- To describe a wave in terms of how it moves in a medium.
- To describe what happens to a medium when two waves try to disturb it at the same place and time.
- To describe the difference between a wave and a moving object.

Materials

- Coil spring

Procedure

One Stretch the coil out on the floor between two members of your group sitting 3 m apart. Practice making wave pulses by quickly moving one end of the coil to the right or left, and then back to the starting point. The other end of the coil should be kept still, and the remaining group member(s) should observe the center of the coil. If you have floor tiles, use the tile edges as a straight line to help observe the motion of the coil. Each group member should have an opportunity to start a wave, to hold the still end of the coil, and to make observations.

FIGURE 4.1
Single pulse.

Two Make a single pulse as seen in Figure 4.1.

Three What happens to the wave pulse as it travels through the coil? What happens when the pulse reaches the end of the coil? Try to repeat this pulse exactly. Are there any differences in how the two pulses moved? Sketch the shape of your wave pulse.

FIGURE 4.2
Full wave.

Four How can you change the amplitude, or height, of a wave pulse? Practice sending wave pulses of different amplitudes through the coil, but move your hand back and forth fast enough to maintain the same period (time it takes for a complete wave to pass a certain point). How do wave pulses of varying amplitude behave differently?

Five Keeping the amplitude the same, change the speed of your hand to make wave pulses with different periods. How do wave pulses of varying periods behave differently?

Six How do the pulse's amplitude and period affect its speed?

Seven Can two waves, or wave pulses, exist in the same place at the same time? Devise a procedure to find out. Carefully observe the center of the coil, and describe your procedure and what you find.

Eight Practice making "full" waves. Quickly move one end of the coil to create waves which have peaks and troughs (see Figure 4.2). Repeat some of the above steps. Do these waves behave the same as wave pulses?

Nine Create periodic waves by repeating the waves in a consistent manner. What do you see? Try different frequencies. Record your observations, and diagram any interesting wave forms you discover.

Ten Explain the wave forms you discovered in Step eight. How does the size of the waves you created compare to the motion required to generate them? What does this mean in terms of energy?

Eleven How are waves different from vehicles in terms of motion?

■ TEACHER-TO-TEACHER

This activity is an opportunity for students to see what they are learning. It is a lot of fun, but does require some supervision and reminders to treat the coils carefully. It will take two periods, and you may want to spend some introductory time discussing waves to determine your student's preconceptions. Students can easily relate to waves in the ocean, yet many may have never thought twice about the unique nature of these waves.

Materials

- Enough coil springs for student groups: Coils should be made of spring steel, and be 2 cm in diameter, 2 m long, and stretch to 12 m. These coils are sold by many educational equipment suppliers for about $10. (These smaller coils will work better than Slinkies.)
- Space for student investigations: A smooth surface is best for this activity. Classrooms, hallways, or a gymnasium will work well. Students should be able to create waves with an amplitude of 50 cm without bumping into another coil.

Teacher's Guide

One No response necessary. You may need to assist students by letting them know if their coil is stretched too far or not far enough, or by helping them initiate wave pulses. Ensure that each student gets to start a wave, hold the still end of the coil, and observe wave patterns.

Two No response necessary.

Three The amplitude of the wave pulse decreases as it travels through the coil. When it reaches the end of the coil, it will return on the other side—if you are looking down at the coil, a pulse that is traveling through the coil on the right, will return on the left. If students can repeat a wave exactly, there will be no differences between the two waves. Diagrams will vary depending on amplitude and period.

Four Amplitude can be changed by moving the beginning of the coil further to one side to initiate a wave pulse. Wave pulses of greater amplitude last longer in the coil, but otherwise behave the same. Short, or long, period wave pulses should behave the same way.

Five There are no actual behavioral differences between long and short wave pulses, although on some surfaces the longer pulses may die out quicker.

Six A wave's amplitude and period should not affect its speed. (See Extension one.)

Seven The two waves will pass through each other, thus existing in the same place at the same time. To observe this, students must initiate a wave pulse from either end of the coil at the same time. They should record that the coil made a different shape when the two wave pulse were on top of each other.

Eight Waves and wave pulses behave the same way. As with Step one, you may need to help some students create waves.

Nine Hopefully students will discover the interference pattern usually called "standing waves." The interference pattern is created by a periodic wave superposing with its own reflection. Diagrams will vary depending on amplitude and period. (See Extension two.)

Ten The amplitude of a standing wave pattern is significantly greater than the wave source. This means the energy of the wave source, your hand, is adding up in the coil.(See Extension four.)

Eleven The speed of a wave can not be controlled by the students, whereas they can control the speed of a vehicle. In addition, unlike waves, two vehicles can not occupy the same space at the same time.

Extensions

Extension one Measure the time it takes a wave pulse to travel a particular length of a stretched coil. The speed of a wave can be calculated from these values. Next, have students control the different variables—amplitude, period, stretched length of coil—and measure their responses in terms of speed. Students should discover that the amplitude and period of the wave pulse has no effect on speed. However, a pulse will travel faster on a coil that is stretched longer.

Extension two Use technology to observe the superposition of waves. Students may have trouble convincing themselves of what actually happens because the "collision" occurs so quickly. You can videotape Step six to definitively observe what happens when two waves "hit" at the center of the coil. The recording will clearly show superposition. You may need several recordings of the "collision" to see all the short times intervals, as the 30 frames per second of standard video could miss some.

Extension three As an introduction to the next two activities, open a discussion about how sound and light travel as waves. As part of this discussion, or as a take-home assignment, explore how different wavelengths mean different sounds, or colors, and how different wavelengths travel in water. One way to start this discussion is with the popular question, "Why is the sky blue?"

Extension four As a take-home assignment or report, ask students how we are able to tune our radios and televisions to one particular station.

Activity 2

Do you hear what I hear?

How does the human ear work?
Can we make a machine that does the same thing?

Challenge

How does your ear transform sound into some signal your brain can interpret? How do other machines transform sound waves into signals that operate something?

Objectives

- To describe the components of the human ear and their functions.
- To describe various technologies that are similar to the human ear.

Materials

- Resource materials for the human ear
- Sound-related devices

Too quiet?

An excellent selling point of many vehicles today is the notion of a silent interior. Who wants to get stuck in traffic and have to hear the jack hammer pounding away right next to their vehicle?

But how safe are sound-proof vehicles on an open road, where the only sound a driver might hear is the siren of an emergency vehicle? In response to this question, EARS Systems, Inc. has developed an Emergency Vehicle Detection System which distinguishes siren sounds from other exterior noises. What kind of technology must be utilized to create this kind of device?

How might this device distinguish the siren of an emergency vehicle from other exterior noises? What other applications can you think of for this kind of technology?

Procedure

One Review the resource materials on the human ear. Label the parts of the ear on Figure 4.3.

Two Describe the overall function of the human ear.

- What does the outer ear do? How is it constructed to perform this function?
- Describe the middle ear and its function. How does it work?
- What is the function of the inner ear? Describe how it works.
- Label the illustration of the ear. Diagram what sound does as it enters and travels through the ear.

Three Which part of the human ear do you think is most remarkable? Why?

Four How does your ear help you operate a vehicle? What advantages and disadvantages do people who are hearing-impaired have when operating a vehicle? How might they compensate?

Five As a group, choose an example of sound-related technology. What does the device do? How? Why?

Six How could your chosen device be used in a vehicle?

Seven What kind of sound-related vehicle technology would you like to invent?

Eight Why does the technology you described in Step seven not exist? Could it? Document your reasoning.

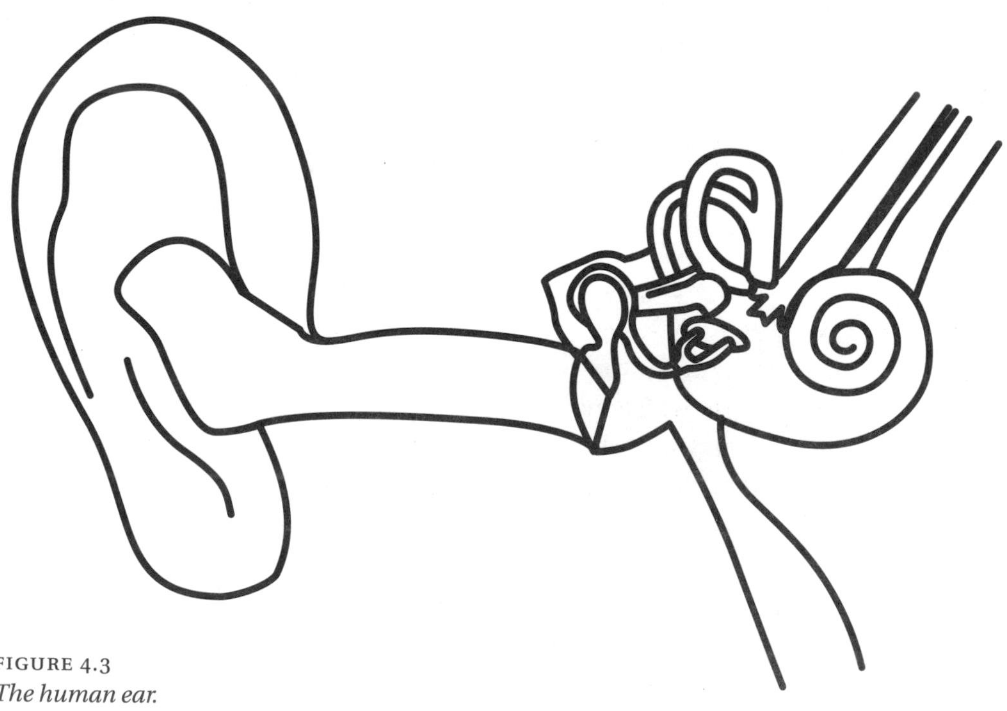

FIGURE 4.3
The human ear.

■ TEACHER-TO-TEACHER

This activity asks students to examine the use of the human ear in terms of its parts and their functions, and also in terms of operating a vehicle. Students will explore the human ear as if it were a piece of technology. They will also investigate existing sound-related technology, then design their own with the goal of enhancing future vehicles.

Background reading for this activity is provided on page 114. You may also need to provide other background resources (standard biology and anatomy textbooks) for the human ear, as well as the technologies you empower students to explore.

Before beginning this activity, you should determine to what depth you want your students to research the human ear. This activity can take one or two periods, depending on this decision.

Materials

- Resource materials for the human ear: A plastic model and written material are best. Possible resources include biology and/or anatomy and physiology texts or medical charts, which may be available from the school nurse. (See "Testing One, Two, Three," page 117.)
- Various sound-related technologies and background information for them: Microphone, speaker, clap-activated on/off switch, old phonograph, or others. Any sound technology accompanied by an operating manual will work. Ideally, you will want different devices for each group. Some students may volunteer to bring something from home.

Teacher's guide

One Ensure students correctly label ear parts.

Two Answers will vary depending upon available resources.

- The outer ear is a sound-gathering device. The ear canal has some resonance functions, like the length of tubing of an instrument.
- The middle ear is responsible for transferring the sound from the air outside the ear to the liquid of the inner ear. Sound can not efficiently refract, or change media (from air to water) so the middle ear mechanically transfers the sound.
- The inner ear is responsible for changing sound to nerve impulses, which the brain interprets as sound.
- Answers will vary depending on available resources.

Three Responses will vary, but students should justify their choice.

Four This question can use any type of vehicle, from skateboard to automobile. The human ear aids vehicle operators by announcing oncoming vehicles by their engine sounds or beeping horns; receiving travel information—weather, traffic, road conditions—from the radio or car phone; identifying mechanical trouble; etc. People who are hearing-impaired may not be able to hear oncoming traffic, emergency sirens, or horns. On the other hand, they will most likely not be distracted by these noises.

Five Descriptions will vary. You may choose to allow students to research their device and provide a formal description, or have them write about what they know and/or can hypothesize.

Six Responses will vary. Encourage realistic answers.

Seven Responses will vary. Again, encourage realism, but also allow for creativity if it is accompanied by an explanation of purpose.

Eight Responses will vary. Students will need to use modern technology resources, such as magazines and the Internet (see page 146), in order to answer this question. This can be used as a take-home activity, or an in-class research project.

Extensions

Extension one Build working models of the middle ear to reinforce the pressure and force transformers involved.

Extension two Have your class investigate their hearing range. A standard lab set-up would include a signal generator with an output meter, and an amplification system that drives a speaker well enough to produce an adequate range of sound. A much simpler and perhaps more powerful method of doing this kind of investigation involves a computer-enhanced lab employing probeware. The repeatability of the sound, as well as the real-time graphing of this technology makes this investigation much more tangible. Students will also gain an appreciation of the difference between low and high frequency sounds.

Extension three As a take-home activity, ask students to research what sounds can cause damage to ears and how. Once they learn the decibel level that is considered to be harmful, have them investigate the decibel levels of various events or places: rock concerts, airplanes and trains—both traveling inside and being outside—a telephone ring, or an average automobile horn.

Extension four How do other animals hear? Assign a different animal to each student, preferably including as many animal classes as possible, as a report. How does the range of an animals' hearing aid it in survival? Some interesting animals include bats, whales, owls, crickets, and ants.

Extension five Have students investigate health or activity-related ear problems that cause hearing loss, including diseases or over-exposure to loud noises. Students could prepare a report, including information on the cause and effect of the problem, how long it will persist, and what, if any, treatments are available.

Activity 3

Picture this

How does the human eye work? How does a camera work? Can our technology see as well as we do?

Challenge

How do your eyes "take pictures?" How do they send images to your brain? Are our eyes like cameras? Do cameras actually see and recognize objects?

Objectives

- To describe the components of the human eye, and their function.
- To compare and contrast a human eye and a camera.
- To investigate visual technologies, and how they assist in operation of a vehicle.

Materials

- Resource materials for the human eye
- Camera
- Visual technologies

Enhanced vision

As you probably know, the accuracy of your vision decreases as the light around you fades. Do you know why? Driving at night can be very dangerous, and driving in the rain even more so. What might you do to increase your visual acuity in the rain?

Car manufacturers in Europe have tackled this and other similar issues related to improving road transport safety. BMW has created the Vision Enhancer System which uses infra-red camera technology to allow a driver to effectively see despite weather conditions.

What is infrared? How can it help you see in the rain? At night? What kind of equipment and technology must be added to a vehicle for this system to work? Where would you put it?

Procedure

One Review the resource material on the human eye available to you. Label the parts of the eye on Figure 4.4.

Two Describe the overall function of the human eye. What are its parts? How does it focus light? How is the image interpreted? Diagram what light does as it enters the eye.

Three Describe the overall function of a simple camera. What are its parts? How does a camera focus light? How is the image stored? Diagram what light does as it enters the camera.

Four How do the cornea and lens create tiny, focused images of the objects we see?

Five How are the focusing processes of an eye and a camera similar? How are they different?

Six How do cameras and eyes control the amount of light they let in?

Seven How does the human eye convert light to nerve impulses?

Eight How do red, green and/or blue cones in the eye react to those particular colors?

Nine How do humans store the images they see? How is this similar to, or different from, a camera?

Ten What type of visual technology is most similar to the human eye and brain? Why?

Eleven How could you combine visual technology with an automobile? Explain.

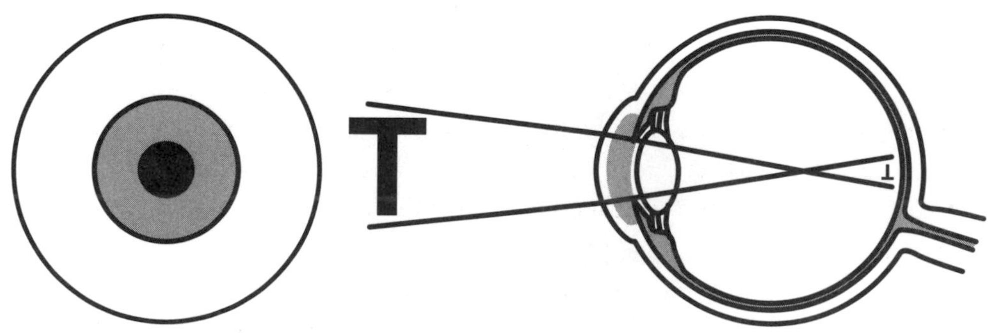

Figure 4.4
The human eye.

■ TEACHER-TO-TEACHER

By comparing and contrasting the human eye to a camera, students will gain a better understanding of how their eyes work. This two-period activity also can be used as an introduction to light, through refraction and lenses, and through how the eye and the camera lens focus light.

Materials

- Resource materials for the human eye: A poster, 3-D model, or reference books. (See "Images and Seeing," page 118.)
- Visual technologies: Ideally something that can be opened for students to explore. Local camera repair stores, pawn shops, or consignment stores, may be willing to donate cameras, or equipment, that is either irreparable, or without value.
- Pen lights (see Extension six)

Teacher's guide

One Ensure students correctly label the parts of the eye.

Two Encourage students to give a more involved response than "to see." The average human eye is convex, and focuses light by refracting it onto the retina. Rods and cones in the retina translate light into nerve impulses which are interpreted by the human brain as the images we see. (See Extensions one and two.)

Three A response of "taking pictures," or "capturing images," can be an acceptable answer. A simple camera uses a convex lens to focus light in a similar fashion to a human eye. More complex cameras use mirrors to reflect images from the main lens to the view finder. The image is stored on film. Diagrams should be based on the equipment available. (See Extension eight.)

Four Students should describe how the cornea and lens "bend" light to produce an image.

Five A camera is similar to the human eye, in that both use a lens system to focus light. The human eye has a fixed distance between the lens and the retina, the focal plane, and therefore automatically adjusts the shape of the lens to create a focused image. A camera, however, has a solid lens, so a photographer must manually adjust the focal plane for the picture to be in focus. (See Extensions three and four.)

Six Both a camera and the human eye have an adjustable aperture, which lets in the correct amount of light. A camera's aperture is the f-stop adjustment, and the eye's aperture is the pupil at the center of its iris. (See Extension five.)

Seven The retina contains specialized cells that act as light receptors. They are called rods and cones. They contain pigments that absorb light and give off nerve impulses. Rods absorb all visible light while cones only absorb particular colors depending upon their pigment: red, green, and blue.

Eight The human eye contains red, green and/or blue cones. Each contains a particular pigment which allows the cone to absorb only that color very well. To see red, for example, the red cones must be stimulated with a bright red light. All the other colors we see are combinations of reds and greens and blues, and we see them as a result of a combination of cones being stimulated. (See Extension six.)

Nine We store the nerve impulses from our eyes in our brain's memory as an image. However, the nerve impulses emanating in the rods and cones are partially processed in the other layers of nerve cells in the retina. Our brains store what we see, but our eye actually "thinks" about the image before the nerve impulses get to our brain.

Ten Responses will vary, however students should understand that there is no existing technology capable of reproducing the human ability to recognize what it is seeing. This question can be used as a take-home assignment.

Eleven Responses will vary. Encourage creativity and inventiveness. This question can also be used as a take-home assignment.

Extensions

Extension one Not all human eyes and brains see and interpret images in the same way. Color blindness, and vision problems—astigmatism, near-sightedness or far-sightedness—arise when some part of the eye is not working in an expected way. Students can research various eye or brain problems which create these conditions, as well as the devices, if any, that help correct them.

Extension two Encourage students to examine their own eyes. They can devise tests to measure their field of vision, determine if they are far or near-sighted, test for color blindness, or anything else they can think of.

Extension three Students can experiment with different types of camera lenses—telephoto, zoom, wide-angle—or even binoculars, for a better understanding of focal lengths and depth of field.

Extension four As a take-home assignment, have students explain how film works, and then compare and contrast it to human memory.

Extension five To demonstrate how the pupil regulates light, have students observe each other closing and opening their eyes. If a student closes his/her eyes for at least 15 seconds, and then opens them, the observing student will be able to see the pupils contract. Dilation, can be observed after the eyes are exposed to a bright light. Have the observing students quickly pass a lighted pen light in front of the other students' eyes. They will see the eyes contract to keep the extra light out, and then dilate again to let the desired amount of light in.

Extension six Explore the biochemical process of light absorption in rods and cones. This process explains many common phenomena—night vision, stroboscopic affects—which students might find very interesting.

Extension seven Ask students to explore how other animals see. Assign a different animal to each student, preferably including as many animal classes as possible. Criteria could include how the animal's eyes, and therefore what it can see, differs from human vision, and how the animal's vision aids it in surviving. Animals with interesting vision systems include flies, sharks, cats, octopuses, and eagles. Students could create oral or written reports of their findings.

Extension eight Have students construct a pinhole camera.

Materials

- Shoe boxes with lids
- Black electrical or duct tape
- One package of enlarging or printing paper: 5 x 7 paper typically comes in 25 sheet boxes that cost about $7 or $8. You may wish to buy an extra box in case one is accidentally exposed to light.
- Paper cutter or scissors
- Stopwatch or watch with a second hand

Procedure

One Use duct tape to tape up any crack through which light might enter the box. Do not tape the lid on at this time. Use a safety pin to make a pin-sized hole in one end of the box. Cover the hole with a 60 mm piece of duct tape.

Two In a dark room, carefully cut a sheet of printing paper to the size of the end of the shoe box. (You may have to cut it a few millimeters smaller). Tape the paper as tightly to the inside of the shoe box as possible, shiny side out, so that it lies flat against the end of the box opposite the pinhole. While you're still in the darkroom, tape the lid tightly onto the shoe box. As before, make sure to cover any cracks through which light might enter the camera. If your school doesn't have a darkroom, you can create one from a windowless room. Hang a dark sheet or blanket across the doorframe, taping it to the wall at regular intervals so that no light shows through. Consider using a red safety light, available (for sale or rent) through most photographic supply stores.

Three Place the camera on a steady, unmoving object, exactly one meter from the object you wish to photograph. The ground will work. You might want to place something on top of the shoe box, to make sure that it does not move. Gently remove the tape from the box. Start timing from the instant the tape is removed. After three minutes, cover the pinhole with the tape. Record the total time elapsed. It is difficult to determine in advance how long the paper should be exposed to light. An exposure time of about three minutes should be sufficient on a relatively sunny day. It might be a good idea to have students expose the paper in different time increments: one student exposes for three minutes, another for two-and-a-half minutes, another for three-and-a-half minutes, and so on.

Four Develop the paper. If you are working with your school's photography class, members of that class could be responsible for developing the paper. If your school does not have a darkroom, you might take the pinhole cameras to a local photography store that does its own printing (rather than sending film out to be processed). Another option might be a photography professor willing to show you how to use a local college's darkroom.

Activity 4

Detection in motion

How do we use our eyes and ears when operating a vehicle?

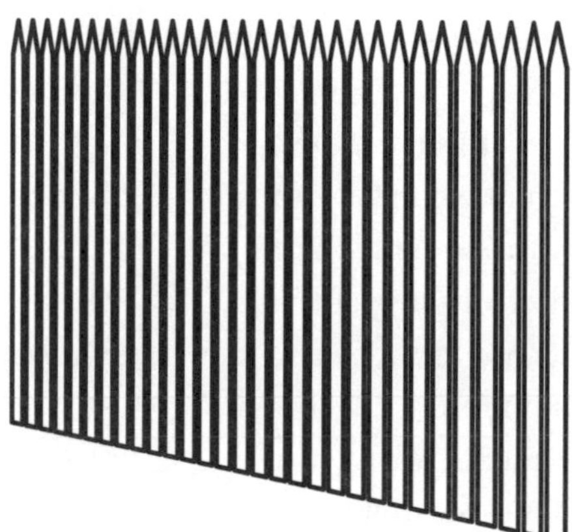

Challenge

Can you imagine riding a bicycle using only your ears for information? Would you know where you were if you traveled in a vehicle with your eyes closed and plugs in your ears? How do we use our senses for gathering information, and how can technology help us along?

Objectives

- To describe how eyes are used when moving.
- To describe how ears are used when moving.
- To create technologies which will enhance hearing and seeing while moving.

Materials

- Group detection packet

Procedure

One As a group, choose a situation from those provided by your instructor. What information does the person in the situation need to know in order to be safe and to get where they are going? How would this person normally get this information?

Two Answer the questions included in your group packet.

Three Present your group's work to the class.

■ TEACHER-TO-TEACHER

Smart cars

What if you had a vehicle that could drive itself? It would be able to "see" the road, "hear" what was going on around it, and always "know" where it is.

Government and private agencies have been working on such vehicles. Intelligent Transportation Systems (ITS) uses computer and communication technologies to address issues from simple mobility to passenger safety.

Another way to achieve a "smart car" is to develop "smart roads." Automated Highway Systems (AHS) complement ITS by adding components to already existing roads that address reaction time and traffic congestion. Vehicles could be placed closer together because an on-board computer would maintain the vehicle's position automatically.

How might you address detection and reaction issues in designing a new vehicle? What kind of technology would you use? Would you focus on the vehicle itself, or the road and the vehicle's interaction with other vehicles on the road?

This activity challenges students to think about transportation in terms of their senses—what they are seeing and hearing when they travel, how they interpret that stimuli, and what their response is. By examining transportation scenarios they are familiar with, students will gain a better appreciation for safety; for example, why they shouldn't listen to music through ear phones while they are riding a bicycle. Students will need at least two periods for this activity.

Materials

- Group Packets: One transportation scenario per group, and one copy of the questions per student (page 113). You can choose any scenario from Chapters 1 and 2 (pages 29–32, and 60–64) for this activity.
- White Construction Paper: One piece per student.
- Markers and/or Crayons: Necessary if Step four is to be done in class.

Teacher's Guide

(Group Detection Packets)

One Chart styles may vary, but a simple grid will suffice. Objects can include anything from other vehicles, to a bump in the road. More than one sensory organ can be included per object. Responses should include only those that would happen within seconds of recognizing the object. For example:

Two Encourage students to brainstorm and write down everything, regardless of how odd they think it is.

Three Students must pick only one device per object.

Four This question can be a take-home exercise. Encourage creativity and inventiveness. Provide each student with a piece of white construction paper, and ask for colored, detailed drawings. Students might also create models of their vehicles.

Five Information about modern technology is available in reference books, magazines, and on the Internet. This question can be the basis of a report, or a simple discussion. Student justification is important, and students should be encouraged to explain why or how the necessary technology would be available.

Extension

Ask your students to research what manufacturers are doing today to provide sense-related, technological, automotive enhancements.

■ GROUP DETECTION PACKET

Procedure

One Create a chart in which your group can list:

- those objects the person in your situation might encounter while in motion
- the sensory organ involved for each and what it does
- the responses the operator might have.

Two What technological enhancements could help the person see, hear, interpret, or respond to the objects in your chart? Work in your group, and record your ideas.

Three Add a column, labeled technological enhancement, to your chart. As a group, determine which of the devices from Step two should be included, and record them appropriately in the new column.

Four By yourself, use the completed chart to create a technologically-enhanced vehicle that incorporates any, or all, of these ideas. Describe your vehicle and include a labeled drawing.

Five Is the technology for your vehicle available today? If not, when do you suppose it might be? Justify your prediction.

Background reading

All about waves

A wave is energy in motion. It cannot be sped up, slowed down, or turned by force. It can neither be stopped, nor held still. So how do you control waves?

To better understand waves, we will start with a definition, and dissect it theoretically.

A wave is a **disturbance** that **moves** through a **medium** from a **source** at a particular **speed**.

A wave is a disturbance In a system in equilibrium, in which the components of the system are all connected to each other and all forces are balanced, disturbing one component would cause forces and motions in surrounding parts of the system. These forces and motions would then be transferred throughout the entire system. If the disturbance was a force acting over a distance, then we could say that work, or energy, had been imparted to the system. The system did not speed up, slow down, or turn as a whole, but the energy was transmitted through the system as the components of the system moved and returned to their original position.

A wave moves Waves are always in motion, which makes it especially difficult to examine them. In most cases, we need specially-designed technology to observe waves and their associated phenomena.

A wave moves through a medium
This may be difficult to envision, as not all waves require a mechanical medium. Sound is a mechanical wave that moves through solids, liquids, and gases quite well. Without a medium, such as in space, there would be no sound. Light, however, consists of vibrating electric and magnetic fields, and these electromagnetic waves require no mechanical medium. Unlike sound waves, light waves, radio waves, and other electromagnetic waves can travel in a vacuum.

A wave moves from a source All waves have a source, and the energy of the wave is the energy from the source. This is easily seen in a mechanical wave. For instance, as you move the end of a coil spring you are doing work on the coil, but the entire coil does not move at the same time. The work or energy is transmitted as a wave pulse. If you applied a greater force, the resulting wave would have a greater energy.

Waves move at a particular speed
The speed of a wave is controlled by its medium. You can change a wave's medi-

um, but you can not apply a force to speed up, stop, or slow down a wave. You can, however, affect a wave's energy. As a source, you can create a high-energy wave (high pitch and/or loud sound) and a low-energy wave (low pitch and/or soft sound), and you can absorb a wave's energy. But, each of those waves has a specific speed in a particular medium. For example, all the different sounds of an orchestra reach the audience at the same time. A particular system—the air in a concert hall—can carry waves at different speeds. The properties of a system can be changed, effectively making it a different medium. A tightly-stretched coil is a less flexible medium, and therefore a different medium than a loose coil. Also, different kinds of waves may travel at different speeds in the same medium.

For example, **longitudinal waves**, where the medium is displaced along the direction of the wave, tend to travel faster in solids than **transverse waves**, where the medium is displaced at right angles to the direction of the wave. This difference is visible in a coil, and can also be seen in nature. In earthquakes, seismic waves, the longitudinal **P–wave** arrives ahead of the transverse **S–wave**. The distance to the source of an earthquake can be calculated based on the difference in time between these two waves.

A wave is different from other objects in motion. Waves are energy that we see as some disturbance in a medium. Waves have size and shape like objects, but wave energy does not take up space like an object. Wave energy has physical characteristics in the medium that we can measure. **Periodic waves**, many waves of the same energy, are commonly measured by **wavelength**, **amplitude**, **frequency**, and/or **period**.

Wavelength is the length of a single wave disturbance, often measured in meters, or meters per cycle. In light, for example, what we see as colors are really different wavelengths of light; red is a wavelength of light that is about twice as long as blue. Amplitude is technically the amount of medium disturbed from its rest position by a mechanical wave, and is also measured in meters. Varying degrees of brightness and loudness are results of different amplitudes of light and sound.

Frequency and period are the time characteristics of a wave. Period is the time for one wave disturbance to pass a particular point, and is measured in seconds per cycle. Frequency is how many wave disturbances pass in one second, and is measured in cycles per second, or Hertz (Hz). Period and frequency are inversely related, the greater the period the less the frequency and visa versa. Also, because a wave's speed is controlled by the medium, a long wave will have fewer disturbances per second—long wavelength, low frequency—and a short wave will have many disturbances pass per second—small wavelength, large frequency. Thus, wavelength and frequency are not independent, whereas an object, such as a long vehicle, can go by a particlar point in any time period depending upon its speed.

The speed of a wave and a vehicle can be measured in the same way—distance divided by time. Unlike a vehicle, however, a wave will always travel the same speed in a particular medium. By measuring wavelength (distance per cycle) and period (time per cycle) speed (distance per time) can be calculated.

Another difference between waves and vehicles is that wave amplitude can change noticeably over distance. Often waves travel outward in many directions, thus the energy from the source spreads over a larger and larger space. This decrease in energy often manifests itself in decreased amplitude, as can be seen in light and sound. The further you are away the source, the light is dimmer and the sound is softer. The energy of a mechanical wave can also be "lost" to its medium. In a coil spring, friction causes some energy to be lost as heat. However, in outer space, sometimes called a "perfect" medium, an electromagnetic wave can travel "forever." Because of this property, light waves from events that occurred billions of years ago can be detected from Earth today.

Waves can bend around objects—a property called **diffraction**—and waves bend most around objects that are about the same size as the length of the wave. The object does not apply a force to the wave, the wave simply bends around it. A sea wave in a harbor can bend around a barrier like a breakwater. This property allows scientists to use electromagnetic waves like x-rays—of which 1,000,000 will fit in a millimeter—to bend around a DNA molecule, enabling them to see patterns from which they can infer structure.

Waves can refract as well, which is what happens when a wave changes mediums. In that instance, it can speed up or slow down, and may even change direction. When sound moves from one medium to another, its wavelength changes, which results in a different pitch. For example, try recording your voice. When you listen to the recording, does it sound like your voice? Because you hear your voice through the medium of your head and not the air, your recorded voice will sound quite different. Now try putting your fingers in your ears. The sounds around you will become muffled because you are interfering with the passage of air into your ears. But, if you speak, you will hear yourself clearly.

Light refracts when it passes through different mediums. Light is also affected by the electric and magnetic fields of atoms. Thus, an electromagnetic wave, like light, traveling through the air behaves differently than when that wave is traveling through glass, water, or other materials. In the 16th century, scientists learned to use this property of light to make glass lenses to focus light.

Not all electromagnetic waves behave like visible light and different substances affect these waves differently. There is a very wide range of energies—wavelengths and frequencies—of electromagnetic waves. Infrared light, for example, has a lower energy and longer wavelength than visible red light. We use specific materials such as glass and special plastics for eyeglasses because they are transparent to visible light and opaque to infrared light. Infrared light lenses, on the other hand, must be made from substances like salt (sodium chloride). Very high energy electromagnetic waves, like x-rays, travel right through our tissues and we can see into ourselves with them.

Like light and sound, electrons also travel in waves. Electrons act like waves in atomic environments, not like tiny "classical" particles. Classical mechanics is the theory that explains how and why things move in our everyday environment. However, it can not explain the motions and energies we measure in environments the size of atoms. In the early part of the 20th century, scientists started to look deeply into how and why very, very small objects move. As they collected that data, they developed the theory called **quantum mechanics** to explain their observations.

Electron microscopes are based upon the principles of quantum mechanics, and use electrons' wave properties—wavelengths smaller than an atom—similar to the way that ordinary microscopes use light. Visible light has a wavelength more than 3,000 times larger than the size of an atom, therefore light is not reflected, refracted, or diffracted by individual atoms. Electrons, however, can have wavelengths smaller than atoms, therefore they interact with objects the size of atoms, molecules, and viruses. We can see our own genes using electrons as waves using electron microscopes.

Testing: One, Two, Three

Our ears are perhaps the most sensitive sensory organ in our bodies. It is a magnificent organ that changes sound waves into nerve impulses that are then carried to our brain for interpretation. Our ears are sensitive to a great range of sound both in terms of loudness (up to 120 decibels, a range of energy from 1 to a trillion) and pitch (15 to 15,000 Hz, a range from 1 to a thousand).

The basic problem our ears solve is how to accurately change sound from compression waves in the air to compression waves in the liquid of the inner ear, and finally to nerve impulses that tell our brains what we hear. To do this our ears have evolved three components: the **outer**, **middle**, and **inner ear**.

The outer ear, the piece we see called the **pinna**, collects sound and channels it into the small **ear canal**. Our middle ear then efficiently transfers those vibrations to comparable liquid vibrations in the inner ear. The **cochlea** in the inner ear, acting as a transducer, changes the sound in its liquid into nerve impulses that are sent to our brains.

The outer ear gives us some ability to locate where sound is coming from—like a radar dish—and has some effect upon the range of sound we hear best. The pinna picks up sound, and by moving our heads we can discover the location of the sound's source. Also, natural frequencies of sound resonate within the ear canal. These frequencies are amplified, like music in the chamber of a guitar, enhancing our ability to hear those specific sounds.

Without the middle ear, the sound would need to travel from the air directly to the liquid in the inner ear. These two mediums have very different properties and much of the sound would reflect back into the air, not refract into the inner ear, thus greatly reducing our ability to hear to about 1/30th of what we can hear.

At the end of the ear canal is the **eardrum** of the middle ear. Sound causes vibrations on the eardrum, which is attached to a tiny bone, the **malleus** (Latin for "hammer"). The sound vibrations are carried as pushes and pulls by this bone. These forces are then transformed through another tiny bone acting as a lever called the **incus** (Latin for "anvil"). The incus then applies a larger force—1.3 times greater—over a shorter distance on the **stapes** (Latin for "stirrup"). The stapes is the bone attached to the membrane between the air space of the middle ear and the liquid of the inner ear. It vibrates that membrane, the oval window, to create sound in the liquid of the inner ear. The surface area of the oval window is much smaller than the eardrum, thereby creating a pressure transformer. The pressure vibrations caused by the sound in the air become about 20 times greater in the liquid of the inner ear. Thus, through the conversion of sound vibrations in air into forces transformed through the middle ear, the sound is almost the same intensity in the air of the outer ear as in the liquid of the inner ear.

In our inner ear, the cochlea acts as a transducer, converting sound into nerve impulses. At this point, sound has become vibrations in the liquid of a chamber in the cochlea. These vibrations cause motions on the membrane between the liquid chamber and the central chamber in which the nerves will be stimulated. There are hairs cells on the membrane, with cilia on their ends, all located in the cochlea's liquid. As the membrane vibrates, it creates sound vibrations in the liquid which moves the cilia of the hair cells. These cilia sway in the liquid much like the hair on your head when swimming underwater. The motion of these cilia produce nerve impulses that our brain can decipher as sound, differentiating pitches and associated loudness. Essentially, we hear sound in our brain.

Images and Seeing: The camera, the eye, and the brain

Visual perception is a fascinating area of study. The human eye forms images of things, processes those images, and then sends nerve impulses to our brains. Our brain interprets the signals, and we see. Sometimes we store what we see in our memory so that we can see it again. Images are so important to us that humans have been developing technology to form and store them throughout history. Recorded history is the result of these imaging technologies, from cave paintings, to carved images, to laser discs.

The human eye employs the phenomenon of refraction in the **cornea** and **lens** to focus light from objects and then form as tiny images on the **retina**. As light passes through the cornea and lens, it slows and changes direction. The shape of these parts of the eye influences this change in direction—negatively, in cases of astigmatism, or myopia. The light converges at a point within the eye and forms a focused image, usually on the retina.

Muscles in our eyes change the shape of the lens, so that we can focus on objects both near and far away. Sometimes, our eyes can not adjust the shape of the lens properly, and blurred, out-of-focus images are formed on our retina. In these cases, we put lenses, either glasses or contacts, in front of our eyes. Lenses correct the problem by altering the direction of light waves before they enter the cornea, which allows our eyes to form clear, focused images on the retina.

The retina is the layer of tissue lining the inside of the back of our eyes. Images are formed here before being sent as nerve impulses to the brain for deciphering. In the retina are specialized cells which convert light into electrochemical signals which travel as nerve impulses. These light sensitive cells are called **rods** and **cones** because of their shapes. Rods are sensitive to any visible light, while cones are sensitive to particular colors, or wavelengths of light: optical red, green, and blue.

The pigment in rods absorbs light within a particular range of frequencies, or wavelengths. Rods convert visible light to electrical pulses—more light, more pulses. Rods then give us the ability to see very dim to very bright light, from gray to white. The range of wavelengths that rods are sensitive to encompasses the visible spectrum.

Different light frequencies, or wavelengths, as well as different combinations of frequencies, produce nerve impulses that our brains interpret as color. There are three different cones, each kind containing a pigment that absorbs and reacts to only one color. Science refers to color based on human visual perception. For example, an object is red because it reflects the wavelength to our eyes that stimulates the cone which our brain associates with red. Black objects give off no light, so our brain receives no color signals, and a white object appears white because it gives off equal red, green, and blue light. All other colors are combinations of red, green, and blue, that our brains "see" as a result of these three signals. Basically, what we think of as color exists only in our brains.

Our brain has a special compartment for interpreting and storing nerve impulses from our eyes, as it does for each of our other senses. We try to build computers to act in similar manners. Our brain is, so far, unique in its ability to quickly recognize a vast array of objects. Today's computers with appropriate transducers can only recognize certain simple objects.

Cameras are somewhat like the human eye. First, light passes through a lens. The lens opens or closes in response to the available light. The shutter opens to allow light into the camera, and the light converges on the film to produce a visible image.

chapter five

Collisions and safety

■ TEACHER-TO-TEACHER

This final learning plan challenges your students to apply what they have learned about force and energy to transportation, while focusing upon safety. It is inquiry-based, and allows your students to delve into questions concerning safety devices they use. Students will learn that the large forces involved in collisions need to be spread out to result in pressures that are tolerable to the human body. In order to understand this, students must consider the energy involved and ways of dissipating it, and they will discover many ways in which the energy of motion can be transferred into less hazardous forms of energy. Finally, students will apply what they've discovered by designing and building a vehicle to protect a "passenger egg," thereby demonstrating they have learned to make scientifically-based, safe decisions about transportation. Students will test their designs in a similar manner to how actual vehicles are tested for safety.

Conceptual development

As forces act upon objects, especially living things, the pressure on the object is more important than the total force, especially when considering the effect on the object other than the resultant change in motion. The following paragraphs describe the concepts developed in the four activities of this chapter.

Energy of motion can be transferred in many ways, and not just through the work of stopping an object and transferring that kinetic energy into heat. Vehicle technology has developed many ways of transferring a vehicle's kinetic energy into some other form without inflicting

Conceptual construction

Prior experiences of collisions combined with knowledge of force and energy while moving in vehicles.

Central Idea Reasonable safety can be designed into vehicles.

- **Collisions, Collisions** Explore various human/animal survived collisions.
- **Spreading Forces** Explore safety devices that lower pressures on humans.
- **Dissipating Energy** Explore safety devices that dissipate energy.
- **Investigating Vehicle Safety** In teams, analyze transportation safety systems.

Constructed & Communicated Idea: Humans can have relatively safe transportation systems.

injury upon the humans involved.

We make decisions about our own safety, and that of others, while moving from place to place. Understanding the science involved in various technologies can establish a foundation from which to make safe decisions.

We all need to be technologically literate in order to make informed decisions. Developing our abilities to design technology, such as safe vehicles, establishes this literacy.

National Science Education Standards

In the development of the conceptual construction of this chapter, the following Content Standard concepts and principles served as a guide.

- Objects change their motion only when a net force is applied. Laws of motion are used to calculate precisely the effects of forces on the motion of objects (page 179).
- Energy can be transferred by collisions ... and many other ways. However, it can never be destroyed (page 180).
- Students should understand the costs and trade-offs of various hazards (page 199).
- Human activity can enhance potential for hazards. Natural and human-induced hazards present the need for humans to access potential danger and risk (page 199).
- Science distinguishes itself from other ways of knowing and from other bodies of knowledge through the use of empirical standards, logical arguments, and skepticism. The core ideas of science such as the conservation of energy or the laws of motion have been subjected to a wide variety of confirmation and are therefore unlikely to change in the areas in which they have been tested (page 201).
- Identify a problem or design opportunity. Propose designs and choose between alternative solutions. Implement a proposed solution. Evaluate the solution and its consequences. Communicate the problem, process, and solution (page 192).

Time management

"Collisions and Safety" should take 10 to 12 fifty-minute periods. Activity 1 will take at least two periods. Activities 2 and 3 will take two to three periods each. Activity 4 is divided into two parts. The time needed for the first will vary depending on how much time your students will need to complete their designs and build their vehicles. The contest in the second part will take at least one period, but you should incorporate at least two others for planning and review, and more depending on whether you do the extensions.

Learning plan assessment

If you've done performance activities like those in the preceding learning plans, your students should now be ready for their presentations to be a major component of their grade for the learning plan.

The poster extension of each group's efforts in Activity 4 can be used as a summary assessment of the entire "Collisions and Safety" learning plan. You will need to create organization and execution criteria in order to prepare your students.

Activity 1

Collisions

How do so many people and animals survive collisions?

Challenge

What enables humans and other animals to survive collisions? How does science help us understand the physical and biological consequences of collisions? Can collisions be defined scientifically?

Objectives

- To describe a collision in terms of the forces involved.
- To describe a collision in terms of energy transfers.
- To describe adaptations and technologies that reduce the risk of injury in a collision.

Materials

- Collision research resources

Procedure

One As a group, pick an animal collision and a human collision. Describe your choices.

Two Sketch the forces acting on the animals and humans during these collisions.

Three Describe the kinetic energy of the animals or humans before the collision. Where does that energy go immediately after the collision?

Four Estimate the area over which the forces acted. What is the effect of spreading the force over that area? Define pressure in your description.

Five What are the variables involved in transferring kinetic energy?

Six How have either the animals or humans spread these collision forces to lower the risk of injury?

Seven How have the animals and humans dissipated the kinetic energy of the collision to lower the risk of injury?

Eight How do so many people and animals survive collisions?

Nine Prepare a brief visual presentation to describe your animal and human collisions. Use the concepts of force, pressure, and energy in your description. Make your presentation as a group.

■ TEACHER-TO-TEACHER

This activity tests any preconceptions students may have about collisions by having them dissect "ordinary" collision situations. Through informal presentations, each student has an opportunity to vocalize their understanding of force, pressure and energy transfers in collisions, and learn how animals have adapted to survive various types of collisions. (See Extension three.)

This activity is a two period, introductory version of the final activity of this chapter (page 137), and is not intended to involve significant research or the preparation of an extensive report. You may want to ask students to choose their scenarios and illustrations before beginning the activity.

Conceptual development

- Collisions involve forces.
- These forces act over areas, doing work to stop a moving object.
- The kinetic energy of the moving object can be transferred into other forms of energy that lower the risk of injury.
- The size of the forces and the area over which they act are one major factor in determining the injury to living things involved in the collision.
- Distance is a major factor in collisions—greater distances result in smaller forces.

Materials

- Resources: Text, Internet, or other media, to research animal adaptations and human technologies that aid collision survival.

Teacher's guide

One You may want to prepare a list of human and animal collisions from which student groups can choose. Ideally, no two groups should analyze the same situation. In addition to obvious collisions, you can also include situations in which animals utilize natural adaptations to survive, and not incur damage from, collisions with the ground.Animal scenarios include: woodpeckers pecking trees, rams butting other rams with their horns, or cats landing on their feet. Human scenarios include: humans boxing, catching a baseball, falling off a bike or other non-motorized vehicle while wearing a helmet, or being involved in any vehicle crash while wearing a safety belt.

Two If necessary, provide force analysis guidelines from "Describing Forces" (page 20). Student descriptions will vary depending upon which collisions they have chosen and their prior understanding of collisions. However, all students should describe the "before" and "after" situations and the contact points of the collisions.

Three If there is no motion immediately after the collision, most of the kinetic energy was probably transferred into heat. If there is a lot of motion right after the collision, such as the shattering of an object into many pieces, much of the kinetic energy may have been transferred into the kinetic energy of the pieces. Also, some of the energy was probably transferred into sound.

Four Estimates do not have to be measured in numbers—they can describe body area, such as all of the torso and the upper part of one arm. By spreading the force over that area, the overall impact is lessened. Students should also recognize that any force acting on a small area causes a large pressure that could cause injury. If necessary, have students read "Standardized Ideas" (page 142) to gain a better understanding of pressure.

Five Variables include work, force, and distance. See "Energy Transfers" (page 80).

Six In both animal and human collisions, biological adaptations and learned behavior help lower the risk of injury. An example of a biological adaptation is the skull, which protects the brain from impact. An example of a learned behavior is that athletes wear large gloves to catch a ball and pull the glove backward as it falls into the glove, thereby increasing the distance over which the force impacts.

Seven This may be a difficult concept for students to identify because the kinetic energy seems to disappear. The heat generated in a collision is biologically insignificant, therefore animals do not seem to have evolved any mechanisms to dissipate kinetic energy. Humans, however, have developed many safety technologies to dissipate kinetic energy and lower the risk of injury. These specific technologies are addressed in "Dissipating Energy," on page 132.

Eight Many animals and humans survive collisions by evolving the means of dissipating force—large ram horns, or complete skulls to protect themselves—or by developing these means through technology.

Nine Develop a presentation rubric for students to follow, and instruct them to prepare their presentation as homework. Each group should competently explain what is happening in their collision, and why the animals and humans involved can survive. Encourage students to be very visual and to have fun, perhaps by safely role-playing through their collision scenarios.

Extensions

Extension one One way to begin this activity is with a collision demonstration. Attach opposite sides of a bed sheet to two shower rods, or two long, narrow pieces of wood (see below). Ask four students to hold up the sheet; two may have to stand on chairs to hold the top so that the bottom does not touch the floor. The two students holding the bottom rod should allow the sheet form a loose pouch on the bottom. Make sure the sheet is loose. Demonstrate the concept of energy dissipation by throwing an egg at the center of the sheet as hard as you can.

Extension two As either extra credit or a required report, ask your students to research one animal's adaptations for collision survival. In this instance, different students can research the same animal.

Activity 2

Spreading forces

How do safety belts and other safety devices protect us?

Challenge

Where does the energy of an impact go? What do we do to lower the risk of impact when we are in motion? How is the force lessened? Why do we sharpen knife blades?

Objectives:

- To investigate the relationships between force, pressure, and surface area.
- To describe collisions in terms of force and pressure.
- To explore the role of pressure as it relates to the sense of touch.

Materials

- Nail boards
- Safety device

Procedure

PART A

One Very carefully, lightly rest the center of the palm of one of your hands on the head of a single nail. What do you feel? Gently press down. How does the feeling change?

Two Without touching the nail board, choose three different parts of your hand, and record a hypothesis for what will happen when you put those parts of your hand on the nail. Why did you choose those three parts? Test your hypotheses. Were you correct? Why, or why not?

Three Create a data table. Record how it feels to rest the palm of your hand on 1, 4, 9, and 16 nails. Do not press down when resting your hand on the nails.

Four What variables are involved in this investigation? How are you controlling them?

Five Add as many columns to your table as variables you identified in Step four, and label them accordingly. Complete your table, using adjectives or numbers to describe or grade the relationship between the number of nails and the variables.

PART B

Six What is the relationship between pressure and surface area?

Seven How do safety devices, such as helmets and safety belts, which are designed to employ this relationship, protect you?

Eight Why are safety belts as wide as they are? Why are they designed to cross your lap and your torso?

■ TEACHER-TO-TEACHER

Caution: This activity deals with materials that can inflict serious injuries. Students must understand that no horse play will be tolerated, and that any student who does anything not described in the procedure will be severely reprimanded.

This activity draws upon Chapters 1 and 3—the basic concepts of motion, force, and energy—to show students how safety belts and helmets work. It is exciting, as well as important, because students will be able to physically feel how force of impact can be "reduced" by surface area. Although the connection between pressure and actual collisions is not overtly stated in this investigation, student data should suggest this connection. You should verbally encourage students to make this connection, and take every opportunity to stress the importance of properly used safety devices. (See Extension one.)

The student procedure section has been divided into two parts, which will take a totla of two to three periods. This allows students to choose their own variables before they are told which variables they should be working with. In order to distribute this activity, cover Part B when you photocopy the page. You should then copy Part B onto a separate piece of paper or ask the questions out loud. You may also wish to combine groups so that fewer nail boards need to be made—more than one student can work with one board at one time. If possible, have students help assemble the nail boards, perhaps for extra credit, either after school or in wood shop.

Conceptual development

- All objects have surface area.
- Forces act over surface area.
- As surface area increases, force "spreads out."
- Pressure, or force per area, is a standard way of describing force.
- In order to examine the effects of force on objects, other than acceleration, it is more important to know pressure than total force.
- To minimize the risk of injury from large forces acting on the human body, some safety devices work by spreading force over large structural areas of the body.

Materials

- Nail boards: Enough of 1″ × 5″ pine shelving to create nailboards for each group. 30 6d 2″ finishing nails per group. Nail the finishing nails into the boards in the following square patterns: 1 nail, 4 nails, 9 nails, and 16 nails (patterns of 36 and 64 nails can be added). Nails should be 1 cm apart, with nail heads even with each other, ½″ above the board (see Figure 5.1).
- Safety devices: Bicycle helmets, elbow and knee pads, football or hockey helmets, automobile safety belts, etc.

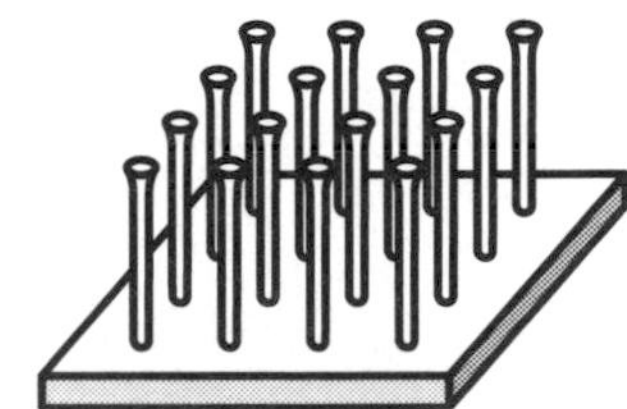

FIGURE 5.1
Nail board.

Students may be able to provide their own. Otherwise, your school's athletic department may be able to help. You may be able to get safety belts at either the school's auto shop department, or a local gas station or garage. Groups can share items if necessary.

Teacher's guide

PART A

One Students will need to take turns, and should be warned against horsing around. Descriptions will very, but all students should recognize that they felt the nail in one place on their palm. If they press a little harder, the feeling should intensify.

Two Responses will vary. Encourage students to choose parts of their hands that are very different—a finger nail, a callused area, the tip of a finger, or a knuckle. Hypotheses may vary as well, and students will find that tougher areas of skin are less sensitive, and therefore feel the nail differently than softer areas. (See Extension two.)

Three Tables can be basic grids (see answer five). Allow students time to progress through the different nail patterns, and compare the differences in feeling.

Four How hard students are pushing with their hands, force; the number of nails on which they are placing their hands, surface area. Introduce and explain these terms if necessary.

Five Adjectives will work well to describe feelings as nail numbers increase. Remind students that they should be pressing down evenly on each group of nails, therefore force is constant for each number. Area can be listed as the number of nails, or the actual area in square centimeters. (See Figure 5.2.)

NAIL BOARD	FEELING	FORCE	AREA
1 nail		1	1
4 nails		1	4
9 nails		1	9
16 nails		1	16
36 nails		1	36
64 nails		1	64

FIGURE 5.2
Sample data chart.

PART B

Six More nails result in a less intense feeling, therefore more surface area means less pressure.
Seven Many safety devices are designed to apply reasonable pressure to the human body. Elbow protectors spread the force of impact over the entire elbow area. Helmets absorb some of the force, and spread the rest over the entire skull. These devices spread the force of the impact, which in turn applies less intense pressure to the body.

Eight Safety belts are wide in order to spread the force of the impact of body against the belt during a stop or collision. Because these collision forces can do severe damage when acting on the "soft" portions of your body like your stomach, they are designed to fit against the bony parts of the body—pelvis and hip region, and rib cage and clavicle.

Extensions

Extension one Use the 9 and 16 nail pattern, a square of Styrofoam that will cover the 16 nail pattern, and several books to demonstrate the concept behind this activity. First place the Styrofoam on the 16 nail pattern, and slowly pile books on top of the Styrofoam, being sure to balance the books, until all the nail heads have been forced into the Styrofoam. Next, turn the Styrofoam over, and place the unmarked side onto the 64 nails. Balance the same books on the Styrofoam. You should find that several books, depending on size, can be balanced on the 64 nails without all the nail heads puncturing the Styrofoam. Allow students to compare the depth to which the nails punctured, or did not puncture, the Styrofoam.

Extension two Ask your students to research how the human sense of touch "works." This can be a report or a short take-home assignment.

Extension three As an in-class extra credit question or take-home assignment, ask students to explain why knives work better when they are sharpened. Students should use what they learned in "Spreading Forces" in their response.

Extension four Have your students make a technical drawing of a safety device for an egg that maximizes the surface a collision force would act over. What shape would it be, and why?

Activity 3

Dissipating energy

What happens if the forces involved in a collision are too great to simply spread out?

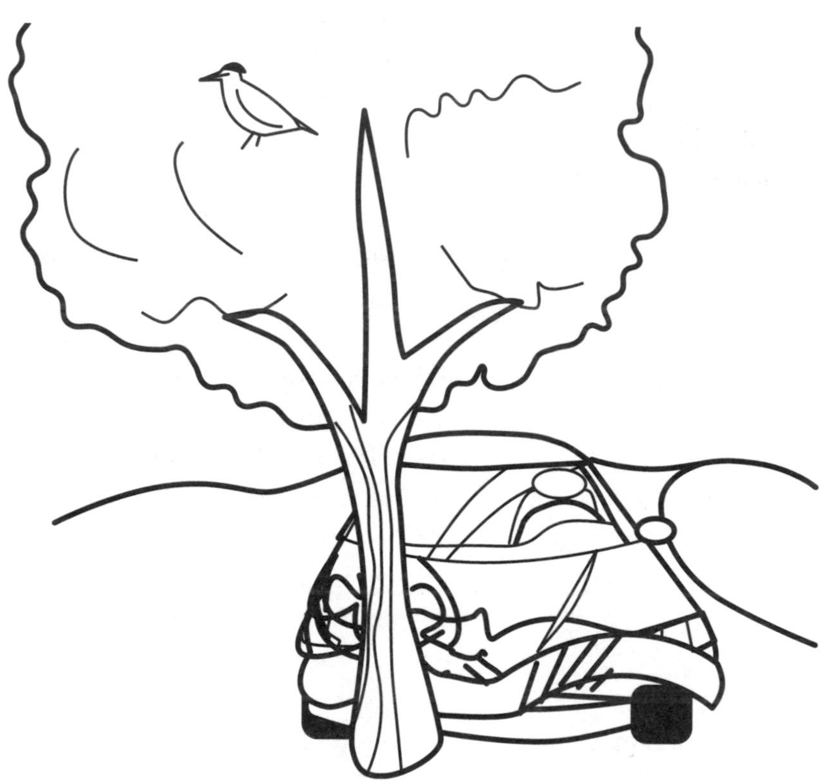

Challenge

Why do the front and back ends of some automobiles “crumple” in collisions? Why do bicycle helmets shatter upon impact? What happens to the kinetic energy in a collision that is not transferred into work while stopping?

Objectives

- To describe energy transfers in collisions.
- To describe energy dissipation in safety devices.

Materials

- Nail board
- Sugar cube brick
- Bread
- Safety goggles
- Cardboard box top
- Safety device

Procedure

One Collect your group's materials. Use caution when handling the nail board.

Two Assemble your sugar cube brick.

Three Organize your group to scientifically observe the event. Record who is doing what.

Four Set up your materials as diagrammed in Figure 5.2. (The crusted side of the bread should be against the nails.)

Five What other data might you want to collect, before you begin the experiment?

Six When your group is ready, check out the hammer from your instructor. Once all group members are wearing their eye protection, hit the sugar cube brick with the hammer. What happens?

Seven What is the "effect" of the sugar cube brick?

Eight Write out an energy analysis of this event.

Nine How are the sugar cube brick, the nail board, and the dinner roll in this experiment similar to an automobile and its passengers?

Ten What happens if the forces involved in a collision are too great to simply spread out?

Eleven Develop a description of how a transportation safety device of your choice dissipates kinetic energy and present that description to your class.

Airbags

Collisions can happen at any speed, but faster moving objects have more kinetic energy and cause more damage in collisions. Scientists have designed bumpers and special materials to help dissipate the energy of the vehicle. But what dissipates energy of the passengers, who are still travelling at the same speed as the vehicle when the vehicle comes to an abrupt stop? The steering wheel? The dashboard? The back of the front seat? The windshield?

Fortunately, scientists have also designed internal safety devices such as safety belts, which both restrain passengers and dissipate energy, and airbags, which solely dissipate energy. In 1998, BMW offered a Head Protection System (HPS), an "inflatable tubular structure," in several of their cars, which protects front seat passengers during side collisions. The HPS also stays inflated for several seconds, offering protection in the event of a roll over or second impact.

How do air bags work? What basic concepts of physics do they incorporate? In 1997, NHTSA announced that people can have the option to disarm their airbags. Why?

FIGURE 5.2
Materials set-up.

■ TEACHER-TO-TEACHER

This activity offers students an inside look at how safety devices and other vehicle adaptations reduce impact by dissipating energy. Students apply an impacting force to simulate a vehicle collision. They will be using potentially dangerous equipment, so all safety precautions should be taken. The activity will take two to three periods.

FIGURE 5.4
Single sugar cube wall.

You could start this activity by showing a video of kinetic energy being transformed in a dramatic way that illustrates how human injury can be reduced in a collision. Possible examples include:

- Hewitt's "Conceptual Physics" Videos (see page 146);
- scenes from the movie "Speed," specifically those which show a vehicle crashing into the water or sand-filled barrels on the freeway; or
- a crash test video showing the front ends of a vehicle collapsing in a collision (see page 146).

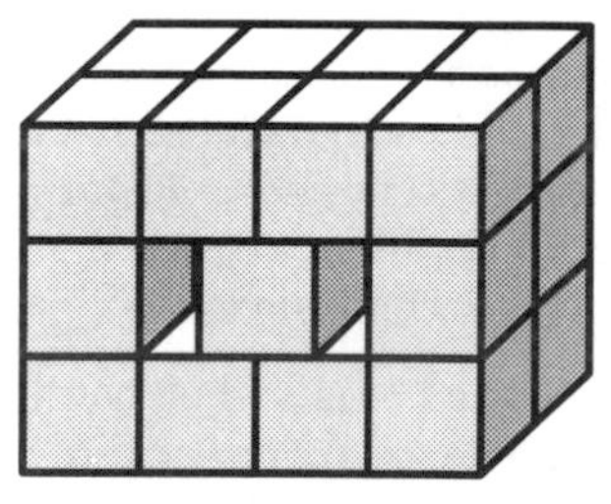

FIGURE 5.5
Single sugar cube brick.

Conceptual development

- Great amounts of kinetic energy are involved when massive objects move, or when any object moves quickly.
- There are ways to dissipate this energy. without relying solely upon kinetic energy transferred by work while stopping.
- Dissipating the energy lowers the risk of being injured by kinetic energy in a collision.
- Scientists have developed technologies that transfer kinetic energy to other objects— such as the pieces of a protective device that shatters or crumples upon impact—or into the energy necessary to physically change some materials.

Materials

- Nail board: One 8 cm × 8 cm piece of ¼″ pressboard, with 16 6d 2″ finishing nails nailed through so that the heads are flush on one side, per group. Nails should be spaced 1 cm apart and arranged in a square.
- 22 sugar cubes per group and white or hot glue: 16 sugar cubes can be used to make two walls of eight cubes each, by gluing them together with white glue (Figure 5.4). The two walls can then be glued together with six other sugar cubes as seen in Figure 5.5.
- Bread: Enough dinner rolls for each group. Rolls should be cut in half lengthwise. Use the bottoms with the crusted side against the nails.
- Hammer or rubber mallet: One hammer or mallet should suffice, and works best for safety reasons. You might wish to have an extra hammer, in case the activity moves too slowly.
- Safety goggles: One pair per student.
- Cardboard box top: The top of a box for copier paper works very well. The box top will contain loose pieces of sugar cube.
- Safety devices for analysis: One device

per group. Devices include: bicycle helmets, wrist guards, knee and elbow pads, and mouthguards.

Teacher's Guide

One Emphasize laboratory safety.

Two If white glue is used, the sugar cube bricks should be allowed to dry overnight. If you decide to have students assemble their own bricks, you should have them do so the day before you begin this activity. In the remaining time, you can show the videos referenced above or review the concepts and procedure included in this activity.

Three Inform your students that in science, the facts of an event must be measurable by more than one scientist to be accepted, and, if possible, the event must be repeatable. Students should organize themselves so that all team members are ready to carefully observe the experiment; they can record their observations after the actual experiment. Depending on your class, you may want to take this time to assign one student from each group to use the hammer.

Four Be certain the teams have adequate space.

Five Students should recognize that they do not know what would happen if they just hit the nail board with the hammer. Without this knowledge, they can not know that performing this experiment without the sugar cube brick will be any different from what they are about to observe. After completing the experiment, have students hit the nail board directly with the hammer, so they can compare their results.

Six No group should be allowed to have a hammer unless all group members are wearing eye protection. Depending on your class, you may want to observe each group as they break the sugar cube brick. The brick should break easily with a good swing of the hammer, and the nail board should not penetrate the dinner roll. (You may want to do this yourself before class, to better explain how students should swing the hammer.) However, the nail board will roll with the hammer swing, resulting in some nail penetration on the hammer side of the roll.

Seven The sugar cube brick protects the bread from "injury" from the nail board by absorbing the energy of the blow.

Eight As a student swings the hammer, chemical energy from his/her body and potential energy from gravity are transferred to the hammer as kinetic energy. The hammer's kinetic energy is transferred to the sugar cube and its pieces, and not into work on the nail board.

Nine This investigation is similar to an automobile as follows:

- the sugar cube brick is like the collapsible parts of the front end of an automobile
- the nail board is like the engine, metal, and plastic that make up the parts of the automobile between the front end and the passenger compartment
- the bread is like the passenger(s)

Ten If the forces involved in a collision are too large to spread out, then much more damage will occur to all objects involved in the collision. In cars, for example, a small collision involving little force might result in a broken head light and/or dented fender. A larger collision resulting from greater speeds or larger objects, and therefore more energy, will involve greater force. This means more damage to the vehicles and potential injuries to passengers. (See Extension three.)

Eleven This question can also be answered as a take-home assignment. Responses will vary depending upon the safety device students chose, but in all cases the energy dissipation factor involves a physical change to an object—a bike helmet shattering or automobile exteriors crumpling.

Extensions

Extension one Some movie stunts and martial art forms take advantage of the ability of energy to be transferred and/or redirected. Students might enjoy learning more about these various energy dissipation and redirection techniques.

Extension two Ask students why larger cars are considered safer than smaller cars.

Extension three Have students write a report on new types of automobile materials used to increase collision safety.

Extension four In 1997, airbags became a controversial subject. As an extra credit project, have students research why some people feel that airbags may do more harm than good. As part of this project, ask your students to compare the data for and against airbags, and draw their own conclusions and solutions.

Activity 4

Investigating vehicle safety

How are safety equipment and devices designed?

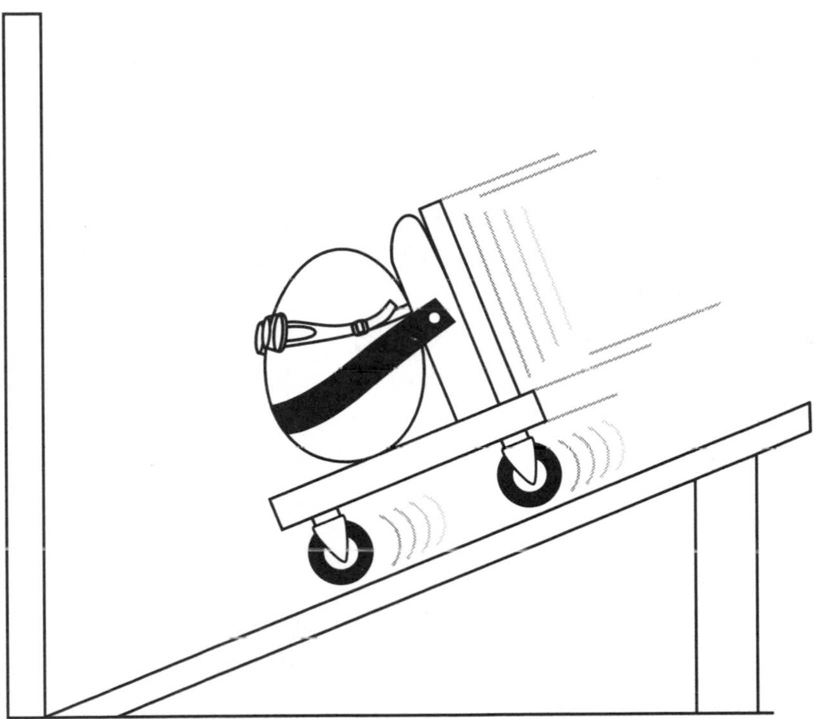

Challenge

Could you make a safety device for a passenger in a vehicle? Do you understand the relationships between force, pressure, surface area, and energy in a collision?

Objectives

- To research, design, build, evaluate, and present safety devices to protect an egg in a collision.
- To demonstrate an understanding of the ideas of force, pressure, and energy in a collision.

Materials

- Vehicle
- Egg in a Ziplock bag
- Safety device building materials

Is bigger still better?

Automobile designers must make compromises between form and function. A bigger vehicle is safer than a smaller one—NHTSA studies have shown that collision fatalities increase by 1.1 percent for each 45.25 kg decrease in vehicle mass—but smaller vehicles have a lot of appeal to a large number of consumers.

In 1997, Ford announced a new bumper design that would reduce the overall weight of the vehicle and still meet the standards for bumper impact.

What parts of an automobile might you redesign to make a vehicle that offers the best possible collision protection? How would you ensure your product is not so massive as to have even more kinetic energy in a crash than the vehicles we drive today?

Procedure

As a group, collect your materials and construct a safe vehicle that will protect an egg during a collision. Assign tasks to group members in order to make efficient use of people and time. Proceed in accordance with the following rules and design process.

Rules

- Safety devices can not cover more than 75 percent of the surface area of the egg.
- No safety device may protrude more than 4 cm in front of the vehicle.
- Vehicles must be turned in before the start of the competition. No modifications can be made to your vehicle after that time.
- Eggs will be provided by your teacher and must remain in the room.
- Each vehicle will be tested at increasing ramp angles. Every vehicle that protects the egg will be approved as a safe vehicle design.

Design Process

One Describe your egg, vehicle, and the collisions it must survive. Record information about the forces, pressures, and kinetic energies involved in the collision between your egg, the safety device you will design, the vehicle, and the wall at the end of the ramp. Explain your descriptions.

Two Generate a design proposal for as many safety devices as you can think of. Write a design brief describing these safety devices and specifying what they are intended to do, in terms of force, pressure, and energy.

Three Devise ways to model and test your ideas without an egg.

Four Develop refined ideas for safety devices by reviewing your research, models, and tests. Review your original design brief.

Five Incorporate your results into a final design. Select the proposed solutions to reduce the risk of injury to your egg from the hazards of force, pressure, and kinetic energy. Why have you made these selections? Have your final design approved by your teacher. Make your vehicle.

Six Record any modifications you made in Step five and the reasons for them.

Seven Before beginning the contest, present your final device to the class. Explain your design, and how it will protect your egg. Record how other final designs differ from yours, and hypothesize what the outcome will be for those devices.

Eight Test your vehicle and record your performance rating.

Nine Based on the results of the vehicle tests, justify the decisions your group made in the design process.

Ten Describe the effects of the design of your safety device. Did your design work as planned? If not, why?

Eleven Evaluate the work of the other groups based on their design and results. Were your hypotheses correct? Why were their designs more, or less, successful?

Twelve Present the results of this design process and communicate your understanding of the concepts of force, pressure, and energy in collisions.

■ TEACHER-TO-TEACHER

This activity invites groups of students to design their own safety device. They will follow a specific design process and apply their understanding of the ideas of force, pressure, and energy, as well as the importance of safety devices. (See Extension one.) By designing and testing a device to ensure the safety of an egg, student understanding can be evaluated equally, even though they may come up with very different designs and have different manufacturing skills. The design process may take one to two periods, and the testing phase will take at least one period.

Conceptual development

- Safety technologies can be developed to spread forces to lessen the pressures on humans during collisions, as well as transforming kinetic energy into less hazardous forms.
- There are limits to the pressures the human body can survive. Through design processes, we can develop safety technologies to greatly reduce the risk of injury during a collision.

Materials

- Ramp: Two to three meters in length, with barrier at base. A flat board held against a wall will work. Ramp will need to be raised and lowered.
- Vehicle: One per student. Vehicles should have a flat bottom and four wheels, but sides are optional. Toy flatbed trucks, or Lego pieces or other construction-styled sets will work. If you use toy trucks or other vehicles, be certain students understand that because the front end of the toy vehicle cannot collapse, it will not dissipate energy.
- Eggs: Enough raw eggs for each group, each in a Ziplock bag.
- Safety device building materials: Including, but not limited to: pieces of egg carton, cotton balls, rubber bands, pieces of balloons, popsicle sticks, plastic straws, tissues, paper clips, paper.
- Protractor

Teacher's Guide

One Descriptions should be accurate. Students should express an understanding of the increased force the egg must withstand, as the increasing ramp angle results in greater speeds, and therefore greater energy. Students may choose to investigate just how much force an egg can withstand by performing their own experiments. You might want to ask students to prepare for this activity by investigating at home, and recording their discoveries in their journal.

Two Depending on available class time, you may want to create a design checklist for your students' design proposals. Each design proposal should include a description of the safety device, what it does and how, and the materials needed for its construction. You should review each group's design proposal to ensure they are working within the reality of the classroom and available materials, and stamp each proposal as "approved." Remind students that there are at least two concepts—pressure and energy transfer—to consider in their designs, especially if they are only focusing on one safety device.

Three Students may not test their initial designs with an egg. Remind them of the dummies used in real vehicle safety tests, and instruct them to be creative and use the observations they made at home as they analyze each of their designs.

Four Make sure students' redesigns are feasible given the available materials and class time. Again, stamp group plan as approved.

Five Students should express an understanding of force, pressure, and kinetic energy in their explanations of their design selections.

Six If students make modifications as they design their final device, they should record their original idea, the modification, and their reasoning.

Seven Before beginning these presentations, collect each group's final vehicle. Remind students that if anyone attempts to make any further modifications, their entire group will be disqualified. These initial presentations should be short, and given by one member of each group. Each group should provide a basic description of the vehicle's safety features, and how they will protect their egg. Allow some time at the end of each presentation for the observing groups to record their hypotheses about their opponent's designs.

Eight Videotape the safety test if possible. Begin at the lowest angle at which the vehicles will roll by themselves. Make sure each vehicle is clearly marked. After each egg breaks, allow students time to record the group to whom it belonged, and the angle at which the egg broke. They should not check their hypotheses from Step seven after each egg has broken, but rather wait until the testing is over.

After testing all vehicles, each vehicle should be rated according to the second to last angle at which the egg survived. For example, a vehicle that performed successfully at 25 degrees, but not at 30 degrees, can only be approved for 25 degrees. Devise a rating system to correspond with angles. One star could be awarded to vehicles which performed successfully between 10 degrees and 19 degrees, two stars for 20 degrees to 29 degrees, etc. (See Extension two.)

Nine Answers will vary. Students should provide clear and concise reasons.

Ten Answers will vary. Students should provide clear and concise descriptions and reasons.

Eleven Answers will vary. Students should provide clear and concise descriptions and reasons

Twelve These presentations should be

longer, and more thorough than those from Step seven. Each group should determine their own format, as well as media. If you were able to videotape the contest, allow the groups to use the footage of their vehicle to explain why their safety device was, or was not, successful. Students should focus only on their design, and present a clear understanding of all factors involved in collisions that have been reviewed in "Collisions and Safety."

Extensions

Extension one Ensure that students do not think the use of safety devices and crumple zones means that they are immune to collisions. Have them research other kinds of injuries that can occur in collisions. For example, a passenger's internal organs also move at the same speed as the vehicle. When the vehicle stops abruptly, as in a collision, there are three impacts: the vehicle collides with an object, the passenger collides with the vehicle, and the passenger's internal organs collide with other organs or the skeletal system. This final type of collision can be fatal, even if injuries from the second impact are not.

Extension two Have students research actual methods of automobile safety testing. Possible sources include Consumer Reports or NHTSA. If your school is located near a facility where automobile safety is tested, arrange for a tour the facility, or for a company representative to speak to your class.

Extension three In addition to the presentations, have your students create posters of their egg and vehicle. They could do presentations for other classes, parents, or students at other schools, to explain collisions, the importance of safety devices, and the reasoning behind how they are engineered.

Background reading

Standardized ideas

If you answer 45 questions correctly on a 50 question test, and your friend answers 90 answers correctly on a 100 point test, who did better? In order to compare these two scores, we must use a **standardized** means of measurement. If you looked at these scores, and determined that 45 is 90 percent of 50, and 90 is 90 percent of 100, then you have used a standardized tool: **percentage**.

Standardized measurements are a tremendous asset to science, and a helpful tool for science students. Very different types of things can be compared and contrasted using standardized measurements. **Pressure** and **density**, for example, are both standardized tools, based on the concept of ratios.

Pressure is the ratio of total force acting over an area. More specifically, pressure is the force that acts on just one unit of area. Right now there is a tremendous amount of air above your body that has weight and is applying a force on you. Although the total force acting on each of us differs according to our size, the atmosphere exerts the same pressure, or force per square meter, on each of us. So, meteorologists and other scientists talk about air pressure, not the total force of the weight of air that acts upon us. The average air pressure at sea level is about 101,300 Newtons per square meter, or 14.7 pounds per square inch (psi). To calculate the total force acting on you, you need to know your total surface area and the current air pressure.

Density is a standardized way of describing mass. Mass takes up space, and density is the measure of how much mass can fit into a certain amount of space, called **volume**. Any one item will always have the same density; a teaspoon of gold has the same density as a ton of gold. Ask yourself this question: If I cut an object in half, does its density change?

Most human bodies have a similar density, even though they may have different masses. This is because we are all made up of the same basic parts—water, bones, tissue—and each part has its own density. Bigger people may have more mass than smaller people, but they also have more space, or volume, inside them to fill. Thus, the density stays relatively the same.

Efficiency is a commonly used standardized tool. It is a way of describing the relationship between input and output, and can be applied in similar situations without necessarily knowing how much of either you may have. For example, if we accept that our bodies are 10 percent efficient, we don't need to know that we ate three Twinkies and ran eight laps to determine that value. Scientists often express efficiency as a percentage. In situations that deal with energy, efficiency will always be less than 1:1 or 100 percent. We can not transform energy from one mechanical form to another without "losing" some to heat.

Specific heat is another standardized measurement. It is the amount of heat necessary to raise the temperature of one cubic centimeter (cm^3) of a material by one degree Celsius (1°C). Because it takes 1 Cal of heat to raise 1 cm^3 of water 1°C, water is the standard to which the specific heat of all other substances are compared. The specific heat of water—1 Cal/kgC$_o$—is relatively high, and most of the solid material of our everyday lives have specific heats less than one, as shown in the chart below.

SUBSTANCE	SPECIFIC HEAT (Cal/kgC$_o$)
Alumninum	0.22
Glass	0.2
Steel	0.11
Wood	0.4
Alcohol	0.58
Human Body	0.83
Water	1

FIGURE 5.6
Specific heat.

■ APPENDIX

	EXCELLENT (3 POINTS)	**OK** (2 POINTS)	**NEEDS WORK** (0 OR 1 POINT)
Research	Does significant outside research from a variety of sources.	Does some limited research.	Does minimal or no research.
Science Application	Applies all pertinent knowledge from curriculum and outside research to decision.	Applies knowledge from science curriculum to decision.	Doesn't demonstrate class knowledge in decision.
Original thought	Demonstrates significant insightful and original thought about the issue.	Completes decision. Gives answers based on logic.	Doesn't think through decision,. Demonstrates no original thought.
Identifying uncertainties	Identifies most uncertainties and suggests research to refine analysis.	Identifies some areas of uncertainty.	Doesn't identify areas of uncertainty.
Analysis	Uses importance bars and sensitivity analysis to analyze decision.	Uses importance bars to analyze decision.	No analysis. Uses gut feelings.
Final decision summary	Makes conclusion using importance bars and expected value. Compares final decision to one not based on analysis.	Makes conclusion based on importance bars. Identifies final decision.	Makes decision, but doesn't explain process.
Writing	Writes persuasively and clearly. Uses logical arguments to summarize position.	Writes clearly. Explains decision situation.	Writes a bit unclearly. Doesn't fully explain the thought process involved in decision making.
Math applications	Uses math to compute expected value and express probabilities.	Uses math to express probabilities.	Doesn't use math.

FIGURE A-1.
Assessment rubric for peer reviews. This assessment rubric can be modified to fit a specific activity.

BIBLIOGRAPHY AND RESOURCES

Bibliography

American Association for the Advancement of Science. *Benchmarks for Science Literacy Project 2061.* New York, NY: Oxford University Press, 1993.

Campbell, Vincent, Jocelyn Lofstrom, and Brian Jerome. *Decisions Based on Science.* Arlington, VA: National Science Teachers Association, 1997.

Flinn Scientific, Inc. *Flinn Chemical & Biological Catalog Reference Manual 1998.* United States of America: Flinn Scientific, Inc., 1998.

Gartrell, Jack E., Jr. *Methods of Motion.* Rev. ed. An Introduction to Mechanics, book 1. Arlington, Virginia: National Science Teachers Association, 1992.

Gartrell, Jack E., Jr. and Larry E. Schafer. *Evidence of Energy.* An Introduction to Mechanics, book 2. Arlington, Virginia: National Science Teachers Association, 1990.

Giancoli, Douglas. Physics: *Principle with Applications.* Upper Saddle River, NJ: Prentice Hall, 1998.

Hewitt, Paul, John Suchocki, and Leslie Hewitt. *Conceptual Physical Science (college).* New York, NY: HarperCollins College Publishers, 1994.

Hewitt, Paul. *Conceptual Physics: The High School Physics Program.* Reading, MA: ScottForseman/Addison Wesley, 1992.

Hewitt, Paul. *Conceptual Physics (college).* Reading, MA: ScottForseman/Addison Wesley, 1989.

Jewett, John, Jr. *Physics Begins with an μ Mysteries, Magic, and Myth.* Needham, MA: Allyn and Bacon, 1994.

Kreighbaum, and Barthels. *Biomechanics: A Qualitative Approach for Studying Human Movement.* New York, NY: Macmillian Publishing Co., 1990.

Lewis, Ricki. *Life.* Dubuque, IA: Wm. C. Brown Publishers, 1995.

Loucks-Horsley, Susan, et al. *Elementary School Science for the 90's.* Andover, MS: Network, Inc., and Alexandria, VA: Association for Supervision & Curriculum Development, 1990.

National Academy of Sciences. *National Science Education Standards.* Washington, DC: National Academy Press, 1996.

Tobin, Kenneth, ed. *The Practice of Constructivism in Science Education.* Washington, DC: American Association for the Advancement of Science. 1993.

Wheeler, Gerald, and Larry Kirkpatrick. *Physics Building a World View.* Upper Saddle River, NJ: Prentice Hall, 1983.

Materials resources

NHTSA Traffic Safety Materials Catalog. To order, contact: Walter Culbreath, (202) 366-1566. Fax: (202) 366-7990.

Internet resources

Government Resources

US Department of Transportation: http://www.dot.gov/

National Highway Traffic Safety Administration: http://www.nhtsa.dot.gov/

Federal Transit Authority: http://www.fta.dot.gov/

NASA: http://www.nasa.gov/

Sandia National Laboratory: http://www.sandia.gov/

University Resources

Machine Vision Based Traffic Surveillance, UC Berkeley: http://www.cs.berkeley.edu/~beymer/traffic-surveill.html

Vehicle Research Institute, Western Washington University: http://www.ac.wwu.edu/~techdept/vri/vri.html

Vehicle Research Laboratory, Delft University of Technology, The Netherlands: http://www-tt.wbmt.tudelft.nl/vehicle/home.htm

Science Education Resources

National Science Teachers Association: http://www.nsta.org/

Science for All Americans: http://project2061.aaas.org/

Roy Beven's Home Page: http://www.smate.wwu.edu/rqb/

Frank Potter's Science Gems: http://www-sci.lib.uci.edu/SEP/SEP.html

Technology and Biomechanics Educational Resources

National Science Olympiad: http://www.

geocities.com/CapeCanaveral/Lab/9699/
International Technology Education Association: http://www.iteawww.org/
TERC: http://www.terc.edu/
Technology for All Americans: http://scholar.lib.vt.edu/TAA/TAA.html
Biomechanics World Wide: http://www.per.ualberta.ca/biomechanics/
Washington Biotechnology Foundation: http://www.wabio.com
Human-Machine System Laboratory in Korea: http://hpsys01.kaist.ac.kr/
Technology Education Index: http://www.technologyindex.com/education/

Technology sidebar bibliography

The following bibliography lists the source of information for the Technology Sidebars found throughout this book.

Less massive metals (page 4)
http://www.thenewsteel.org
http://www.ulsab.org
For additional information about the Ultralight Steel Autobody, contact: Mr. E. Opbroek, ULSAB Program Director, ULSAB, 2000 Town Center, Suite 1900, Southfield, MI 48075-1138, EdOpbroek@ULSAB.org, (248) 351-1757.

Have force, will travel (page 21)
http://www.zapbikes.com/home.html
For additional information, contact: ZAP Power Systems, 117 Morris St., Sebastopol, CA 95472, (707) 824-4510.

No friction? No problem (page 45)
http://www.asha.corp
For additional information, contact: Elliot Goldberg, ASHA Corp., 600C Ward Drive, Santa Barbara, CA 93111, (888) 345-2742.

Alternative fuels (page 68)
http://www.ott.doe.gov/program/oaat/afv.html
http://www.ethanolrfa.org
For additional information about alternative fuels, contact: John Garbak, U.S. Department of Energy, 1000 Independence Avenue, SW, Washington, DC 20585-0121. John.Garbak@hq.doe.gov, (202) 586-1723.
For additional information about ethanol, contact: Renewable Fuels Association, One Massachusetts Avenue, NW, Suite 820, Washington, DC 20001, etohrfa@erols.com, (202) 289-3835.

Thinking and driving (page 69)
http://pecan.srv.cs.cmu.edu/afs/cs/user/pomerlea/www/ralph.html
For additional information, contact: Dean A. Pomerleau, a Research Scientist at Carnegie Mellon University, at pomerleau@cs.cmu.edu.

Fuel efficiency (page 81)
http://www.ott.doe.gov/program/oaat/hev.html
For additional information, contact: Rogelio A. Sullivan, U.S. Department of Energy, 1000 Independence Avenue, SW, Washington, DC 20585, (202) 586-8042.

Too quiet (page 102)
http://www.ears-na.com
EARS System Inc.
For additional information, contact: Customer Service, 311 Blue Ridge Place, Ballwin, MO 63011-2435, infousa@ears-na.com, (800) 550-3277.

Enhanced vision (page 106)
http://www.bmw.ca.
For additional information, contact: BMW of North America, P.O. Box 1227, Westwood, NJ 07675-1227, (201) 307-4000.

Smart cars (page 112)
http://www.its.dot.gov
http://www.its.dot.gov/ahs/
For additional information, contact: Raymond Resendes, U.S. Department of Transportation, 400 Seventh St., SW, Washington, DC 20590, raymond.resendes@fhwa.dot.gov, (202) 366-2182.

Airbags (page 133)
http://www.bmwusa.com
For mo additional information about BMW's Head Protection System, contact: Robert Mitchell, BMW of North America Inc., P.O. Box 1227, Westwood, NJ 07675, robert 514@aol.com, (201) 307-3790.

Is bigger still better? (page 138)
http://www.ford.com
For additional information about Ford's light-weight bumper system, contact: Sara Tatchio, Ford Motor Co., Ford World Headquarters Room 922, Dearborn, MI 48121, statchio.ford@e-mail.com, (313) 323-8116.

Several of the above were located through "Auto Product NEWS," an Internet publication of the Society of Automotive Engineers, at http://www.elecpubs.sae.org/APN/frames.